# SEMI-ENCLOSED SEAS

## Exchange of Environmental Experiences between Mediterranean and Caribbean Countries

T0179133

To commemorate

the Quincentenary of the Discovery of the Americas

Published with the collaboration of
ENTE COLOMBO '92, Genoa, Italy

# SEMI-ENCLOSED SEAS

## Exchange of Environmental Experiences between Mediterranean and Caribbean Countries

*Edited by*

## PAOLO FABBRI

*University of Bologna,*
*Italy*

and

## GIULIANO FIERRO

*University of Genoa,*
*Italy*

Routledge
Taylor & Francis Group

LONDON AND NEW YORK

1992

First published 1992 by Elsevier Science Publishers Ltd

2 Park Square, Milton Park, Abingdon, Oxfordshire OX14 4RN
52 Vanderbilt Avenue, New York, NY 10017

*Routledge is an imprint of the Taylor & Francis Group, an informa business*

First issued in paperback 2019

WITH 30 TABLES AND 14 ILLUSTRATIONS

**British Library Cataloguing in Publication Data**

Semi-Enclosed Seas: Exchange of
Environmental Experiences between
Mediterranean and Caribbean Countries
  I. Fabbri, Paolo   II. Fierro, Giuliano
333.9

  ISBN 978-1-85166-849-6 (hbk)
  ISBN 978-0-367-86553-5 (pbk)

**Library of Congress CIP data applied for**

# FOREWORD

The Renaissance inaugurated an age of new frontiers: the New World as a reflection of a magnified European reality.

The debate over India following Columbus's expedition is based on the consideration of the capacity of nature to support innovations introduced by existing technologies.

Nowadays, the environment appears to be a closed system in which resources are limited and frequently non-renewable. As a consequence, even the rights to exploit these resources could be limited. In the balance between man and nature, the individual takes second place to the survival of humanity.

This hierarchy dictates the destiny of future generations, since decisions taken today carry long-term environmental consequences.

The law of nations makes allowances for a Law of the Sea in order to balance individual initiatives with the common good.

The available knowledge and level of technical development are unequally distributed. Thus, there can be no guarantee of unity between economic and environmental policies.

Reality changes, as everybody knows. It is up to people, states and large international organizations to decide upon the direction and the aims of any such development.

The celebration of the fifth centenary of the great navigator Christopher Columbus gives us the opportunity to rethink the responsibilities with which the 'new future' of the world is faced.

The sea has always represented a source of knowledge and a frontier of research. Demographic pressure, the search for new forms of income, simple curiosity and the development of nautical instruments have stimulated and allowed exploration of the seas.

It is the sea that has carried the great explorers and colonizers to their goals. The first wave of colonizers were motivated solely by their courage. The second wave consisted of sponsored expeditions under the aegis of the great powers of Europe, in order to extend and reinforce their control over the new lands.

Most of the time, Italian navigators served foreign sovereigns and did not participate in the conquest of the New World. The Italians currently describe events and point out the environmental differences between America and Europe, underlining the necessity for an approach combining both ways of thinking, thus allowing the establishment of mutual, collaborative agendas.

The technical evolution of shipbuilding and of nautical instrumentation places a greater responsibility on the shoulders of joint ventures between states, bonding their external relations and thus guaranteeing the respect of agreements drafted between them.

In the interest of the common good, today's world needs vigilant supervision through the creation of laws ruling the protection and use of the seas and the natural environment, without compromising the conditions necessary for economic survival and development.

The protection of the seas is at the forefront of today's environmental concerns.

IVO BUTINI
*Deputy Minister*
*for Foreign Affairs*

# PREFACE

Over the last decades, semi-enclosed seas and their coastal zones have been faced with a notable increase in population and often uncontrolled and uncoordinated processes of development. Such processes have occurred in all maritime countries regardless of their level of development, although along different lines. They have resulted in coastal pollution, irrational use of resources and deterioration of living conditions.

Ten years of personal experience within the Mediterranean Action Plan have shown me how useful and in fact necessary can be cooperation between countries faced with common problems on issues of environmental protection. Comparative analysis between diverse local and regional situations, involving complex and multidisciplinary problem assessments, is likely to lead to proper methodologies in problem-solving, which could not otherwise be pursued through sectoral investigations.

Further, my experience within the field of coastal geology on issues of littoral dynamics and shore protection has led me to the conclusion that such studies should be carried through the widest possible range of approaches.

On these grounds, and thanks to the precious support of Ente Colombo—for which I should like to express my personal gratitude to Alberto Bemporad and Giorgio Doria—it has been possible to call a meeting in Genoa, in such an important year for the city, of the riparian countries of two of the most important and populated semi-enclosed seas, the Mediterranean and the Caribbean.

The achievement of common guidelines for both of these areas is stimulating further action with the assistance of already existing supranational bodies, such as the European Community for the Mediterranean and Caricom and FAO for the Caribbean.

The opportunity offered by the Genoa International Exhibition and by other conferences held within this framework has stimulated the proposal for a meeting between the Action Plans of distant areas, but still similar in physical as well as cultural features. It was felt that this was a case where an exchange of ideas and experiences could be particularly valuable.

This was the first opportunity for a joint meeting between the experts working on the two Plans, and one may hope that it has resulted in a successful mutual check on the validity of the methodologies adopted.

Although quite different in such oceanographic features as water circulation, the two basins do show similarities, especially in the area of principal interaction between man and the sea, defined as the coastal zone. In this area, issues of environmental protection and management look particularly crucial in semi-enclosed basins. More than half of the world's population presently lives within 60 km of the shoreline, and future projections conducted by UNEP indicate that within the next 20–30 years this proportion is likely to increase significantly, resulting in further environmental degradation and over-exploitation of marine resources.

Three issues believed to be of special relevance were proposed for discussion at the meeting:

• *Water quality*, including pollution from terrestrial sources and from oil discharge at sea. In both basins, any major oil leak implies a potential situation of high risk, as was the case of the 'Haven', which was discussed at the meeting. To prevent these disasters, an oil combating centre (REMPEC) has been established in Malta.

• *Coastal zone management*, including coastal planning as a device to face ever-increasing coastal development, the impact of tourism and the preservation of cultural heritage.

• *Sea-level rise*, which could lead in the near future to dramatic erosion and flooding processes—an issue on which further information has to be acquired at a global level.

I should like to express my thanks to the Italian government, which was represented at the meeting by the Deputy Minister for Foreign Affairs, the Hon. Ivo Butini, and other senior officers of the Ministries of Foreign Affairs and of the Environment. Special thanks are due to the 32 representatives of the interested countries—17 from the Mediterranean and 15 from the Caribbean area—as well as to representatives of the EEC, UNEP, the Caribbean Environmental Programme and the Action Plans Coordinators, Ms B. A. Miller for the Caribbean and Mr S. Busuttil for the Mediterranean.

I shall close with the hope that the exchange of ideas and profitable discussions, which have led to closer ties between the two Plans, may lead to more systematic cooperation and to the implementation of the initiatives included in the Final Recommendations, which were unanimously approved.

GIULIANO FIERRO
*Scientific Co-ordinator*
*of the International Meeting*

# CONTENTS

# Contents

# INTERNATIONAL MEETING ON UNEP REGIONAL PROGRAMMES IN THE MEDITERRANEAN AND CARIBBEAN SEAS
## Genoa, Italy, 12–14 February 1992

## PARTICIPATING COUNTRIES

| *Caribbean Action Plan* | *Mediterranean Action Plan* |
|---|---|
| Bahamas | Albania |
| Barbados | Algeria |
| Colombia | Cyprus |
| Costa Rica | EEC |
| Dominican Republic | Egypt |
| Grenada | France |
| Guadeloupe | Greece |
| Guatemala | Israel |
| Honduras | Italy |
| Jamaica | Lebanon |
| Mexico | Libya |
| Netherlands Antilles | Malta |
| St Lucia | Monaco |
| Suriname | Morocco |
| Turks and Caicos Islands | Spain |
| USA | Turkey |
| | Yugoslavia |

UNEP/Ocean and Coastal Areas Programme Activity Centre
UNEP/Regional Coordinating Unit for the Caribbean Environment Programme
UNEP/MAP Coordinating Unit for the Mediterranean Action Plan
UNEP/MAP Regional Activity Centre for the Priority Action Programme (RAC/PAP)
UNEP/MAP Regional Activity Centre for the Priority Action Programme (RAC/SPA)
UNEP/MAP Regional Activity Centre for the Blue Plan (RAC/BP)
UNEP/IMO/MAP Regional Marine Pollution Emergency Response Centre for the Mediterranean Sea (REMPEC)
Atelier du Patrimoine de la Ville de Marseille Réseau 100 Sites Historiques

# THE REGIONAL PROGRAMME ON ASSESSMENT AND CONTROL OF MARINE POLLUTION (CEPPOL) OF THE CARIBBEAN ENVIRONMENT PROGRAMME

Ms. Beverly A. Miller
Acting Co-ordinator
Caribbean Environment Programme

The programme in the Wider Caribbean that deals with water quality is called "Regional Programme for the Assessment and Control of Marine Pollution" (CEPPOL). The CEPPOL Programme is a joint programme between the Intergovernemental and Oceanographic Commission of UNESCO and the Caribbean Environment Programme and UNEP. In 1989 a regional workshop to review priorities for marine pollution monitoring, research, control and abatement was held in San Josè, Costa Rica. This workshop established the joint UNEP/IOC CEPPOL Programme.

The components of this programme are: Programme coordination, control of domestic, industrial and agricultural sources of pollution; baseline, studies on pesticide contamination and formulation of control measures; the monitoring of the sanitary quality of bathing and shellfish growing waters; the monitoring and control of pollution by oil and marine debris; site specific studies of damaged ecosystems and the development of proposals for remedial action; the development of environmental quality criteria and effluent guidelines and finally an activity on the significance of organotin as a pollutant in the Wider Caribbean. The question might be raised as to why organotin. In light of the dependency of the region on tourism, it is important to determine whether organotin contamination represents a serious threat to marine ecosystems in the Wider Caribbean and to recommend the appropriate control measures. All of this is due to the heavy pleasure boat activity in the region. Why do we need a CEPPOL programme in the Caribbean?

The marine and coastal environment of the Caribbean Sea and the Gulf of Mexico provides a major source of wealth, directly or indirectly and supports a large number of people.

The regions near shore, coastal resources, beaches and coral reefs are the basis of industry and supply the regions fishery resources.

It is very readily recognized that the balance of payments crisis is one of the major problems facing a number of countries of the region. If we can cut down on our import bill this will certainly help the region. The countries need to extract for the Caribbean near shore environment, as many economical benefits as it possible.

What are the sources of marine pollution in the Caribbean. Throughout the region pollutants introduced by man are degrading and destroying the marine and near shore areas. The major sources are domestic and industrial wastes. At present the region is at various stages of industrial development. Consequently, the amounts and levels of these wastes will vary depending on the level of industrialization of the particular coastal area involved. You know, I mentioned earlier, this morning, the whole maritime aspect of the Wider Caribbean region, the waters of Caribbean region, are potentially one of the largest oil-producing areas in the world. The petroleum industry alone generates 70% of Venezuela's national income and is critical to the economies of Trinidad and Tobago, Mexico and the Gulf States of the United States.

In addition to oil production, a steady stream of tanker traffic moves approximately 5 million barrels of petroleum trhough the Panama Canal. The Caribbean region has on of the world's heaviest tanker movements through narrow channels, and in the vicinity of small Island increasing in this manner some ports the possibility of shipping accidents.

What has been the impact of this marine pollution? Marine mammals, fish, and shellfish have been severely impacted by hydrocarbon pollutants and I know that the participants from the Mediterranean are aware of the impacts of this type of pollution.

During last year (1991), the objectives of the CEPPOL programme were to organize and carry out the regionally coordinated marine pollution monitoring and research programme. As explained by a previous speaker, there is the need to enter into agreements with the various laboratories and to come up with analytical methodologies and approach as that ensures quality and comparability of the results. Another objective is to generate information on sources, levels, amounts, trends and effects of marine pollution to formulate proposals for technical administrative and legal pollution control measure. At the regional level existing legislation includes the Cartagena Convention and its protocols. However, that does not mean very much in terms of national applicability, one still has to move the regional legislation to the national level and this can be achieved through the CEPPOL Programme. This permits decisions makers the opportunity to develop national legislation with the use of results originating from monitoring programmes. I believe that one can say that the major source of marine pollution in this region is domestic wastes. Sewage is the number one water pollutant and the number one marine pollution problem in the Caribbean. It is interesting to look at the outcome and the recommendations of the CEPPOL Seminar on Monitoring and Control of Sanitary Quality of Bathing and Shellfish-Growing Marine Waters in the Wider Caribbean, which was held in Kingston, Jamaica, 8-12 April 1991.

The meeting recommended that the following be considered as priority activities within CEPPOL, taking into consideration the limited funding available:

1. Those countries, which have already adopted EEC, WHO or pre 1986 US/EPA standards or guidelines for the bacteriological quality of bathing and recreational water quality, or who have developed their own, should continue to use them until sufficient information is available to select more appropriate water quality criteria indicator organisms for the region.

2. For those countries in which the above criteria do not exist, two alternative interim measures are recommended:

    a) the adoption of a faecal coliform criterion based on the recommendations of either the EEC, WHO or US/EPA pre 1986.

    b) the adoption of enterococci indicator organisms as recommended by the EPA 1986 criteria, using s selected health risk factor.

3. To examine the relationships among faecal coliforms, enterococci, E. coli and faecal streptococci concentrations in tropical waters and the survival rates of these organisms. Research should be conducted into the potential use of other indicator organisms, such as F+ coliphages.

4. Countries should develop adequate monitoring programmes to assess the state of faecal pollution of bathing waters as well as its sources. Those

programmes should be based on a common methodology and standardized procedures for sampling, analysis, statistical interpretation and evaluation for the above parameters. Sampling and analysis for faecal coliforms is the recommended minimum activity. The recommended methodology is contained in the Regional Seas series of Reference Methods for Marine Pollution Studies.

5.  Baseline studies should be conducted to determine the sampling frequency, time and depth, as well as the siting of sampling points.

6.  To minimize faecal contamination of bathing waters, treatment and disposal technologies that are reliable and require simple operation and minimal maintenance (e.g. land treatment, waste stabilization ponds and long sea outfalls) should be promoted and supported whenever possible.

7.  As faecal pollution of bathing waters is not an isolated problem, a holistic approach should be adopted in the search for and implementation of solutions of the problem of wastewater treatment and disposal.

8.  Epidemiological studies should be conducted to determine the relationship between the above mentioned indicators and health risk from recreational water contact. While the Seminar recognizes the difficulty and the expenses involved in conducting such studies, the valuable information collected would contribute towards the establishment of regional health-based criteria.

9.  The countries of the region should begin to evaluate their health statistics to ascertain what relative public health risk associated with primary contact recreation in coastal waters would be acceptable.

10. For those countries in the region with shellfish industries the Seminar recommends the adoption of either the USFDA or EEC guidelines for shellfish harvest water quality as interim measures.

11. Literature reviews should be conducted to determine the appropriateness and feasibility of alternative sewage treatment technologies such as the small bore

(diameter) sewer system and the anaerobic upflow reactor, among others, as wastewater collection and treatment alternatives.

12.  Re-use sewage effluent and reduction of wastewater generation should be promoted, especially where fresh water resources are limited and stressed. To this end, a literature review should be prepared on the advantages, implications and available methods for water saving and for wastewater reduction and reuse.

13.  Additional technical and/or financial support should be provided to the national institutions, in particular to those with the least infrastructure, in order to ensure a wider participation of laboratories in the previously recommended studies.

I would like to make the point that the re-use of waste water is a typical issue in the Wider Caribbean region. A number of countries are of the view that it is impracticable and unacceptable to re-use sewage effluent as a source of drinking water. However, a number of countries are considering the treated sewage effluent for agriculture purposes. Water as a resource diminishing in the region. Clean water is in great demand. The entire re-cycling and re-use scenario, with respect to water is now being seriously considered. The CEPPOL programme is new, it has identified sewage as the main water quality problem, the standards for the region are yet to be developed, we have had not too many problems in terms of the standards that are being used within the region, such as ECC, WHO and US/EPA.

The Wider Caribbean region can be marketed as an area where it is still possible to find clean water sources. The threat of marine pollution is on the rise and the Caribbean Environment Programme through the marine activities of the CEPPOL Programme is actively moving towards diminishing threats from tourism, industry and the ever increasing population growth.

With respect to the national Statistics, it is unfortunate that our colleague from Cuba Ms. Salabarria was unable to present you with this type of information. However I am hopeful that this brief intervention somehow will assist you to understand the water quality situation in the Wider Caribbean Region.

Mediterranean Water Quality:   The Mediterranean Action Plan and its
Environmental Assessment Component

Francesco Saverio Civili
Marine Scientist
Co-ordinating Unit for the Mediterranean Action Plan
United Nations Environment Programme
Leof. Vassileos Konstantinou, 48, Athens 116 35, Greece

## ABSTRACT

The paper first recalls the major events which led to the adoption of the
Barcelona Convention, the launching of the Mediterranean Action Plan and
its environmental assessment component (the MED POL programme).   The
programme is described in detail including information on the monitoring
component, supporting activities (provision of equipment, organization of
training, intercalibration and data quality assurance programme) as well
as the research component with its five study areas.   In the second part
the paper makes an attempt to cover the major marine environmental
pollution problems of the Mediterranean.   Levels of the most important
Mediterranean pollutants are given in sea-waters, sediments and biota, as
available.   The data mentioned in the paper are the results of the first
ten years of MED POL.   In its conclusions, the paper briefly recalls the
major achievements and major problems of the MED POL programme after
fifteen years of implementation.

## INTRODUCTION

As early as the time of the first settlements man realized that rivers -
used downstream for drinking - could not be used as a sink; on the
contrary, the seas have always been considered a very effective sink for
any type of waste.   This is unfortunately particularly evident for the
Mediterranean which has been for thirty centuries a crossroad of
civilizations and the means for international trade and industrial
development.   Considering the high rate of evaporation of the
Mediterranean Sea in comparison to precipitation and land run-off, this
basin acts as a negative estuary and this negative water balance requires
an inflow through the Straits of Gibraltar of about 1,7 million $m^3$/sec of
waters from the Atlantic Ocean.   The estimated renewal time of the
Mediterranean waters is between 80 and 100 years [1].   As a result, this
relative high turnover rate allows for an enormous waste receiving
capacity of the Mediterranean Sea;   however, in more recent times its
receiving capacity has been superseded in many areas and the resulting
pollution is having an ever increasing impact on the social and economic
well-being of the population bordering it.   To make it worse, the
horizontal circulation patterns of the Mediterranean and the complex
structure of the water column cause restricted circulation in many zones
of the basin in which pollutants can get trapped for long periods.   In
addition, large scale circulation may transport pollutants from one area
to another, thus spreading concern equally to all bordering States and,
because of the touristic vocation of the Mediterranean, even to people

from other far countries who choose the Mediterranean as their
recreational site.

River run-off, although relatively small (of about 0.014 million
m$^3$/sec), contributes to large pollution loads mainly in highly populated
and industrialized areas, especially in the northern Mediterranean.

Domestic and industrial waste-waters are also discharged in large
amounts, largely untreated, through the sewerage systems of the cities
and towns that line up almost continually in long stretches of shoreline.
In fact, of the 350 million inhabitants of the 18 Mediterranean Coastal
States, more than 130 million live on the narrow strips of coastal plain
and on the hillsides bordering the sea (fig.1).

The amount of organic matter discharged into the coastal zone by
far over-runs the small natural productivity, giving rise more and more
often to severe eutrophication problems, extremely difficult to deal
with.

Nutrients, heavy metals, petroleum hydrocarbons and high molecular
weight chlorinated hydrocarbons are discharged from land-based sources in
amounts that affect, at times catastrophically, living organisms.

Shipping across the Mediterranean Sea, mainly of oil being produced
along the east and south shores and consumed in the north and west
shores, is another source of pollution reaching alarming levels mostly in
the southern Mediterranean.

Yet, in the Mediterranean the demand for using the marine and
coastal environment is one of the largest, if not the largest, of the
world's oceans. Power plants (conventional and nuclear), refineries and
petrochemical industries are competing with each other and with the more
traditional activities, such as fishing or tourism, at such social costs
than can be found nowhere else.

## THE MEDITERRANEAN ACTION PLAN AND ITS ENVIRONMENTAL ASSESSMENT COMPONENT (MED POL)

The deteriorating state of the Mediterranean Sea started to provide cause
for concern approximately two decades ago when adverse health effects on
bathers started to be directly linked to seawater quality - which in
several coastal recreational beaches was evidently not up to acceptable
standards - and when shellfish grown in the vicinity of urban
agglomerations had become suspect. At the same time, evidence of the
disappearance of natural marine life, or damage to coastal ecosystems,
mounted. Since then efforts have been made by Mediterranean states and
by a number of international organizations to halt the progress of
pollution and to rehabilitate as many coastal areas as possible.
However, it appeared evident that such action needed first of all a
comprehensive survey of the state of pollution of the Mediterranean Sea,
the identification of the causes and the acquisition of a better
knowledge of the particular characteristics of the Mediterranean.

As a result, in 1972 the Food and Agricultural Organization (FAO),
the International Commission for the Scientific Exploration of the
Mediterranean Sea (ICSEM) and the Intergovernmental Oceanographic
Commission (IOC) prepared the first comprehensive review of the state of
marine pollution in the Mediterranean.

As a follow-up, in 1974, the General Fisheries Council of the
Mediterranean (GFCM) of FAO, organized two consultation meetings on the
protection of living resources and fisheries from pollution in the
Mediterranean and in January 1975, UNEP convened in Barcelona an
Intergovernmental Meeting attended by 16 coastal States and the European
Economic Community (EEC), at which the Mediterranean Action Plan was
adopted and measures were taken for its implementation [2].

One year after the Meeting, UNEP convened the Conference of
Plenipotentiaires of the Coastal States of the Mediterranean Region at
which the Barcelona Convention and the first two Protocols were adopted.

The Mediterranean Action Plan, as approved in 1975, includes three
basic components: Environmental Legislation, Environmental Management
and Environmental Assessment.

All the components of the Action Plan are meant to be
interdependent and provide a framework for comprehensive action to

8

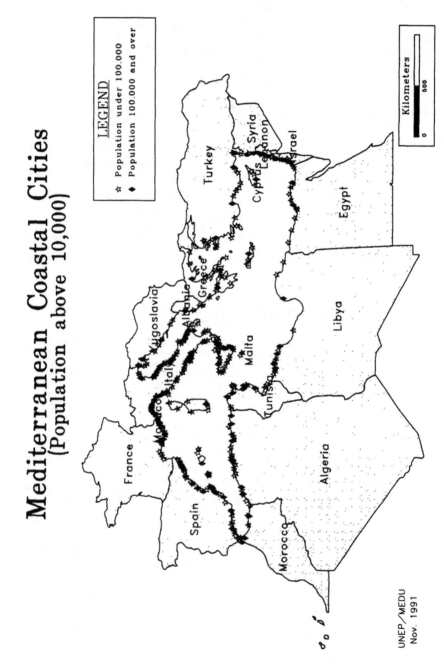

# Mediterranean Coastal Cities
## (Population above 10,000)

LEGEND
☆ Population under 100.000
◆ Population 100.000 and over

Kilometers
0    500

France
Spain
Morocco
Italy
Yugoslavia
Albania
Greece
Turkey
Syria
Lebanon
Cyprus
Israel
Egypt
Libya
Tunisia
Malta
Algeria
Morocco

UNEP/MEDU
Nov. 1991

Figure 1. Mediterranean Coastal Cities (Population above 10,000)

promote both the protection and the continued development of the Mediterranean region. No component is an end in itself. Each activity is intended to assist the Mediterranean Governments in improving the quality of the environmental information on which the formulation of their national development policies are based. Each activity is also intended to improve the ability of Governments to better identify options for alternative patterns of development, and make environmentally sound choices for allocation of resources.

Since 1975, various intergovernmental meetings have taken place, expanding the scope and contents of the Mediterranean Action Plan to a programme of about six million US dollars per year (1991). A relatively complex network of institutions, responsible for its implementation, is co-ordinated by UNEP acting as secretariat to the Barcelona Convention through the Co-ordinating Unit for the Mediterranean Action Plan, which was established by the decision of the First Meeting of the Contracting Parties and is based in Athens.

The environmental assessment component of the Mediterranean Action Plan was designed to provide continuous information on the actual state of pollution of the Mediterranean Sea and to provide the necessary inputs towards the preparation, adoption and updating of regional legal instruments, as well as the formulation of national legal and administrative measures to prevent and control pollution.

The initial pilot phase of this component, termed the Joint Co-ordinated Pollution Monitoring and Research Programme in the Mediterranean (MED POL - PHASE I), was carried out between 1975 and 1981 and was followed, in 1981, by the Long-term Programme for Pollution Monitoring and Research in the Mediterranean Sea (MED POL - PHASE II) which initially covered a ten year programme and was later extended to 1995 to enable the full fulfilment of the objectives of the Programme [3].

The MED POL - PHASE I programme was co-ordinated by the Regional Seas Programme Activity Centre of UNEP until 1981 when it was handed over to the newly created Co-ordinating Unit for the Mediterranean Action Plan. Day-to-Day co-ordination of the activities was carried out by FAO, WHO, IOC, WMO and IAEA. A number of other agencies, namely, UNESCO, UNIDO, etc., collaborated in some specific aspects of the programme.

More than 200 scientific groups of 84 institutions from 16 countries bordering the Mediterranean participated in MED POL - PHASE I activities. The data produced by the research centres were submitted to the agencies responsible for the day-to-day co-ordination, and validated, taking into account the results of the intercalibration exercises co-ordinated by the International Atomic Energy Agency (IAEA) from its International Laboratory for Marine Radioactivity (ILMR), at Monaco, and were processed by the Mediterranean Co-ordinating Unit, in co-operation with the relevant Co-operating Agency.

## MED POL - PHASE II

The Programme's pilot phase activities that started in 1975 came to their end in 1980. The experience gained through them allowed UNEP and the various Co-operating Agencies (FAO, WHO, IOC, UNESCO, WMO and IAEA) to define a Long-term Programme for Pollution Monitoring and Research in the Mediterranean (MED POL - PHASE II), which was adopted by the Second Meeting of the Contracting Parties to the Barcelona Convention [3].

The second phase of the MED POL will last until 1995 when a new programme covering monitoring and research activities will be proposed to governments for approval. The activities are carried out by national laboratories designated by the Mediterranean Governments and co-ordinated by UNEP's Co-ordinating Unit for the Mediterranean Action Plan, in close co-operation with the six Co-operating Agencies mentioned above.

A standing Scientific and Technical Committee composed of MED POL National Co-ordinators was also established by the Contracting Parties to assist them in their review of the progress of the programme and the evaluation of the results. The Committee's task is also to advise UNEP on technical and policy matters related to the programme and prepare recommendations for submission through UNEP, as the secretariat to the

Convention, to the meetings of Contracting Parties.
The objectives of this long-term programme are designed to provide, on a continuous basis, the Contracting Parties to the Barcelona Convention with:

- information required for the implementation of the Convention and Protocols;
- indicators and evaluation of the effectiveness of the pollution prevention measures taken;
- scientific information which may lead to eventual revisions and amendments of the relevant provisions of the Convention and the Protocols, and for the formulation of additional Protocols;
- information which could be used in formulating environmentally-sound national, bilateral or multilateral management decisions essential for the continuous socio-economic development of the Mediterranean region on a sustainable basis; and
- periodic assessment of the state of pollution of the Mediterranean Sea.

The MED POL - PHASE II programme is divided into two different groups of activities involving, respectively, monitoring and research, which are being implemented following different procedures.
The monitoring activities are organized through National Monitoring Programmes, which include a workplan for the monitoring of the sources and of the coastal and reference areas clearly stating geographic boundaries of the areas monitored. They also contain information on sampling sites and frequency, analytical techniques, equipment available and its current state, sea-going facilities, institutional arrangements and any other relevant information, including financial and human resources available as well as the requirements for additional financial and technical assistance, or training.
The research activities are based on individual research proposals submitted by national research centres officially designated by the National Co-ordinators for MED POL. The proposals are scientifically analysed by the relevant Co-operating Agency and the financial requirements are assessed before a research agreement between the agency and the national research centre is established for a period between one and three years.

**The Monitoring Activities**
Four types of monitoring are being carried out:

Monitoring of sources: The purpose of monitoring the sources of pollution is to provide information on the type and amount of pollutants reaching the marine environment from coastal sources, to establish the pollution load reaching the Mediterranean Sea and to contribute to the understanding of biogeochemical cycles of pollutants relevant to the Mediterranean Sea. It covers the:

- survey of the type and amount of pollutants discharged directly into the coastal waters from Land-Based (coastal) Sources;
- survey of the type and amount of pollutants dumped directly into the sea;
- survey of the type and amount of pollutants dumped in emergency or released accidentally into the sea; and
- assessment of the type and amount of selected substances reaching the sea directly through natural (weathering, hydrothermal, etc.) processes from Land-Based (coastal) or Marine Sources.

This monitoring is based on reports submitted by the Contracting Parties according to Article 7, Article 8 and Article 9 of the Dumping Protocol; Article 8 and Article 9 of the Emergency Protocol; Article 6 and Article 13 of the Land-based Sources Protocol; as well as reports submitted by the Contracting Parties on monitoring of sources for substances which may contribute substantially to the overall level (concentration) of pollutants in the sea and data generated by the monitoring of effluents. Parameters covered by this type of monitoring are listed in table 1.

TABLE 1
Parameters to be determined in effluents

---

The parameters to be monitored are divided into two categories:

Category I:

- Parameters which should be included in national monitoring programmes within the framework of MED POL.

Category II:

- Parameters which should be included in national monitoring programmes whenever necessary and applicable.

Category I parameters:

Volume and characteristics of discharge (e.g. pH, temperature, general composition).

- Total mercury (HGT)
- Total cadmium (CD)
- Total suspended solids (TSS)
- Total phosphorus (P)
- Total nitrogen (N)
- Faecal coliformes (FC)
- BOD/COD
- High molecular weight halogenated hydrocarbons (HH)

Category II parameters:

- Petroleum hydrocarbons (PHC)
- Detergents (DET)
- Phenols (PHE)
- Total chromium (CR)
- Selected radionuclides (RAD)
- Other pollutants known to be discharged in significant quantities

Note: In the specific case of monitoring industrial effluents, the parameters listed above and/or other parameters should be selected in accordance with the specific composition of the wastewater discharged.

---

Monitoring of coastal areas: The main purpose of monitoring the coastal areas, including estuaries, within the limits defined by Article 1 of the Barcelona Convention and by Article 3 of the Land-based Sources Protocol, under the direct influence of pollutants from identifiable primary (e.g. outfalls, discharges or coastal dumping points) or secondary (rivers and other water courses) sources, is to establish the effects of measures taken by Contracting Parties under the Land-based Sources Protocol (Article 8(b). Table 2 indicates the matrices and parameters covered by this type of monitoring.

Monitoring of reference areas: The purpose of the monitoring of reference areas, which are not under direct influence of pollutants from identifiable primary or secondary sources as defined by Article 1 of the Convention, is to provide information on the general trends in the level of pollutants in the Mediterranean Sea.
The monitoring of areas falling within national jurisdiction is based on the work of governmental selected national research centres, while the monitoring of areas outside of national jurisdiction is agreed jointly by the governments concerned.

TABLE 2
Parameters and matrices selected for monitoring of coastal areas,
estuaries and reference areas

### Monitoring of coastal waters

Category I parameters:

- Total mercury in organisms and sediments (HGT)
- Organic mercury in organisms (HGO)
- Cadmium in organisms and sediments (CD)
- High molecular weight halogenated hydrocarbons in organisms and sediments (HH)
- Faecal coliforms in recreational waters and bivalves (FC)

Category II parameters:

- Basic oceanographic and meteorological parameters SP (e.g. salinity, oxygen, temperature, chlorophyll, wind)
- Floating tar balls and tar balls on beaches (TR)
- Total arsenic in organisms (AS)
- Radionuclides in organisms (RAD)
- Pathogenic microorganisms (PAT)
- Polynuclear aromatic hydrocarbons in organisms (PAH)

Note: Other parameters could be included according to local requirements.

### Monitoring of estuaries

Category I parameters:

- Total mercury in organisms and sediments
- Organic mercury in organisms
- Total cadmium in organisms and sediments
- High molecular weight halogenated hydrocarbons in organisms and sediments
- Faecal coliforms in water and bivalves
- Total phosphorus in water and suspended matter (N)
- Total nitrogen in water and suspended matter (P)
- Total suspended matter (TSS)
- COD
- Basic oceanographical and meteorological parameters (salinity, oxygen, temperature, chlorophyll)

Category II parameters:

- Radionuclides in organisms
- Polynuclear aromatic hydrocarbons in organisms
- Phenols in water (PHE)

TABLE 2 (Cont'd./..2)
Parameters and matrices selected for monitoring of coastal areas,
estuaries and reference areas

---

**Recommended species for monitoring** (BIOTA)

a) Bivalves

   Mytilus galloprovincialis (MG), or
   Mytilus edulis (ME), or
   Perna perna (PP), or
   Donax trunculus (DT)
   M. edulis, P. perna or D. trunculus can
   only be monitored as alternative species if
   Mytilus galloprovincialis does not
   occur in the area.

b) Demersal fish

   Mullus barbatus (MB), or
   Mullus surmuletus (MS), or
   Upeneus molluccensis (UM)

   M. surmuletus or U. molluccensis can only
   be monitored as alternative species if
   Mullus barbatus does not occur in the
   area.

c) Pelagic carnivore fish

   Thunnus thynnus (TT), or
   Thunnus alalunga (TA), or
   Xiphias gladius (XG)

d) Pelagic plankton feeding fish

   Sardina pilchardus (SP)
   Other clupeids should only be
   monitored as alternative species if
   S. pilchardus does not occur in the
   area.

e) Shrimps

   Parapenaeus longirostris (PL), or
   Nephrops norvegicus (NN), or
   Penaeus kerathurus (PK)

   N. norvegicus or P. kerathurus can only be
   monitored as alternative species
   if P. longirostris does not occur in
   the area.

**Monitoring of reference areas**

The same parameters (both Category I and Category II selected for
coastal waters.

Selection of reference areas is made taking into account the present knowledge of the prevailing conditions and other relevant regional programmes in the Mediterranean Sea. Parameters and matrices covered by this type of monitoring are basically those for the coastal areas (table 2), with the exception of micro-organisms.

Monitoring of the transport of pollutants to the Mediterranean Sea through the atmosphere: The purpose of this monitoring, as defined by Article 4 of the Protocol for the protection of the Mediterranean Sea against pollution from Land-based Sources, is to establish the input (flux) of pollutants into the Mediterranean Sea through the atmosphere and thus to provide additional information on the pollution load reaching the Mediterranean Sea.

The monitoring is based on the work of national research centres designated by their governments.

The monitoring areas include (i) areas directly influenced by identifiable sources of air pollution and (ii) reference areas not directly influenced by identifiable sources of air pollution.

Monitoring of areas outside of national jurisdiction, or under shared jurisdiction by two States, is agreed jointly by the Governments concerned.

Parameters (indicators) to be monitored are selected on the basis of their relevance to the annex I and annex II of the Land-based Sources Protocol.

**The Research Activities**
Only research topics directly relevant to the achievement of the objectives of MED POL - PHASE II are envisaged. The topics included until 1988 in the MED POL - PHASE II programme were very broad and were covering the development of sampling and analytical techniques, the development of reporting formats, the formulation of the scientific rationale for the environmental quality criteria, epidemiological studies, the development of proposals for guidelines and criteria governing the application of the Land-based Sources Protocol, research on oceanographic processes, research on the toxicity, persistence, bioaccumulation, carcinogenicity and mutagenicity of selected substances listed in annexes of the Protocols, research on the eutrophication, the study of ecosystem modifications, the effects of thermal discharges, the biogeochemical cycle of specific pollutants and the study of pollutant-transfer processes.

Since 1989, in order to bring the research component even closer to the initial objectives of MED POL and to the implementation of the Protocol on pollution from Land-Based Sources, the initial research activities (a to l) were refocused and divided in five research areas as follows:

Research area I - Characterization and measurement: This area includes projects which cover the characterization (identification of chemical or microbiological components) and measurement development and testing of methodologies of specified contaminants;

Research area II - Transport and dispersion: This area includes projects which aim at improving the understanding of the physical, chemical and biological mechanisms that transport potential pollutants from their sources to their ultimate repositories. Typical topics are atmospheric transport and deposition, water movements and mixing, transport of contaminants by sedimentation and their incorporation in biogeochemical cycles. Priority is given to the provision of quantitative information ultimately useful for modelling the system and contributing to regional assessments;

Research area III - Effects: This area includes projects relevant to the effects of selected contaminants, listed in Annexes I and II of the LBS and Dumping protocols, to marine organisms, communities and ecosystems or man and human populations. Priority is given to effects and techniques providing information useful for establishing environmental quality criteria;

Research area IV - Fates/Environmental transformation: This area includes projects studying the fate of contaminants (including microorganisms) in the marine environment such as persistence or survival, degradation, transformation, bioaccumulation etc., but excluding transport and dispersion which is dealt in area II;

Research area V - Prevention and control: This area includes projects dealing with the determination of the factors affecting the efficiency of waste treatment and disposal methods under specific local conditions as well as the development of environmental quality criteria and common measures for pollution abatement.

## Supporting Activities

Analytical techniques, intercalibration and data quality assurance: Sampling and analytical techniques used in the monitoring are based on Reference Methods issued by UNEP in co-operation with the relevant UN Agencies, after consultation with and testing by, scientists participating in the programme. At present seventy Reference Methods have been published or are shortly to be issued [4]. Other methods are in principle accepted, including remote sensing, subject to a satisfactory intercomparison. In order to ensure the highest degree of quality, and of comparability of data, all national research centres participate in the continuing intercalibration of sampling and analytical techniques, which is an integral part of MED POL - PHASE II. The national research centres participating in monitoring receive standards and reference substances enabling them to compare their performance with that of other Mediterranean and non-Mediterranean laboratories. Weaknesses detected through this quality control programme are corrected through training and technical assistance, whenever necessary.

In 1988 a more comprehensive Data Quality Assurance (DQA) Programme was initiated in cooperation with IAEA, with the purpose of creating closer and more effective contacts with the scientists involved in the analyses and facilitating possible interventions on the analytical procedures and the instruments. In fact, the new Data Quality Assurance programme includes visits of experts to assess the state and the potential of the selected laboratories, discussions with scientists on the philosophy and the strategy of the monitoring programme, assessment of the capabilities of the scientists and consequently of the training needs, and visits of the MED POL electronic engineer to check the state of the instruments. Also, an expert carry out together with the local scientists the actual sampling and analyses identifying possible mistakes and corrective measures. An internal reference material is also prepared for the laboratory and, at the same time, a split sampling is carried out to enable the expert to control the quality of the analyses after his departure.

This programme, initiated in a number of countries from the South of the Mediterranean, is proven very efficient especially in the cases of developing countries with limited experience, and it is expected to continue and develop in parallel to the more traditional intercalibration exercises.

Data analysis and dissemination: Data reported through the National Co-ordinators for MED POL, in the case of monitoring activities, or through the Co-operation Agencies in the case of research activities, are subject to a preliminary quality control and analysis, in close co-operation with the national research centres, or other organizations collecting them.

Depending on their nature, the data are reported to UNEP's Co-ordinating Unit for the Mediterranean Action Plan in agreed formats and according to agreed schedules. A second analysis of data is carried out, including the control of their quality, in view of the laboratory's performance in the Intercalibration or the Data Quality Assurance programmes, as applicable.

Groups of experts are convened by the Co-ordinating Unit, and Co-operating Agencies, to assist in the analysis, integration and interpretation of data.

Data are computerized at UNEP's Co-ordinating Unit and introduced in the MED POL Data Bank. Reports are then prepared by UNEP on the basis of the data gathered from the various sources and are submitted to the Contracting Parties after they are reviewed by the Scientific and Technical Committee.

The type of documents which are prepared as output of the programme are:

- Reports on the type and amount of pollutants directly entering the Mediterranean Sea from Land-based Sources.
- Results of selected research and study topics.
- Periodically updated reports on the state of pollution of the Mediterranean Sea, indicating the major environment problems, general trends in the pollution of the Mediterranean, as well as the environmental problem which may face the Mediterranean basin in the future.

In addition, the following reports/publications are produced:

- Directories of marine research centres and scientists;
- Bibliographies of marine pollution;
- Reference Methods for marine pollution studies;
- Guidelines for pollution control;
- Assessments of the state of pollution by individual pollutants;
- Environmental quality criteria and standards;
- Emission standards and standards of use;
- Scientific reports; and
- Progress reports, etc.

**Direct assistance**: In addition to the assistance envisaged through the distribution of documentation, reference samples and standards, direct assistance related to MED POL - PHASE II is provided, as follows:

- Individual and collective training for scientists and technicians in methods required for their effective participation in monitoring and research takes place in the form of fellowships, visits of experts, attendance at workshops, seminars, and study tours, etc;
- Analytical equipment and standardized material etc. are provided to national centres participating in MED POL - PHASE II to enable their full participation in the monitoring and/or research programme;
- A common maintenance service for the analytical equipment (atomic absorption spectrophotometers, gas chromatographs, etc.) is at the disposal of laboratories participating in the programme, enabling the proper functioning of their equipment.


## STATE OF POLLUTION

This chapter represents a summary of the results of the first ten years of MED POL. The information which are reported are based on the results of the monitoring and research activities carried out by the laboratories who have been participating in MED POL and are integrated with the information already existing on the Mediterranean. The content of this chapter is largely taken from document "State of the Mediterranean Marine Environment" MAP Technical Reports Series No. 28 published by UNEP in 1989 [5].

**Mercury**
Mercury is of special importance for the Mediterranean. Several countries bordering the Mediterranean have laws which set a limit for the Hg-Total concentration in seafood and, as will be shown below, many fish and shellfish species caught in the Mediterranean exceed this limit. Therefore, if the laws were enforced, large parts of the catch would be withheld from the market. For example, France has a legal limit of 0.5 mg Hg-T kg$^{-1}$ FW and Italy 0.7 mg Hg-T kg$^{-1}$ FW. The high intake of mercury

through seafood is reflected in higher Hg levels in the hair and blood of persons eating considerable amounts of seafood than of those eating little or no seafood. Therefore, the high Hg levels in certain Mediterranean seafood species present a legal and possibly a sanitary protection problem in addition to any effects these levels may have on marine organisms and ecosystems.

Sea water: Sea water concentrations vary over a wide range and the use of different methods without intercomparison or intercalibration with a sea water standard makes comparison of the data difficult, if not impossible. Older data, i.e. prior to 1980, report means up to 120 ug Hg-T $l^{-1}$ and ranges even higher, but recent data have also high means: total dissolved Hg may reach 8 ug $l^{-1}$ and reactive Hg or Hg determined with anodic stripping voltametry can result in means of 10 ng $l^{-1}$. Levels of various Hg forms from other oceans are in general lower than the Hg concentrations observed in the Mediterranean, but the accuracy of the data cannot be ascertained until quality control with standards or intercalibrations has been carried out (Table 3).

TABLE 3
Selected mercury concentrations (ng $l^{-1}$) in sea water from the Mediterranean and other regions.
(extract from MAP Technical Reports Series No. 28, table 15, Annex)

|  | n | mean | range | location | sampling depth | reference |
|---|---|---|---|---|---|---|
| Mediterranean | | | | | | |
| open sea: | | | | | | |
| Hg-T | 3 | 92 | 62 - 110 | Gibraltar | 15 - 300 | Robertson et al., 1972 |
| Hg-T | 47 | 10 M | 5 - 17 | NW Medit. | 25 - 2500 | Huynh-Ngoc and Fukai 1979 |
| Hg-Td | 4 | 25 | 20 - 30 | Tyrrhenian | 0 - 5 | Fukai and Huynh-Ngoc 1976 |
| Hg-Td | 54 | 7.2 | 1.4 - 19.2 | Tyrrhenian | 0 | Ferrara et al., 1986 |
| Hg-T | 2 | 120 | 90 - 140 | Cyprus | 15 - 300 | Robertson et al., 1972 |
| Hg-A | 6 | 30 | 5 - 80 | Ionian-Centr. | 0 - 5 | Huynh-Ngoc and Fukai 1979 |
| Hg-A | 3 | 40 | 15 - 80 | Aegean | 0 - 5 | Huynh-Ngoc and Fukai 1979 |
| Hg-A | 4 | 16 | 12 - 20 | S. Levantine | 0 - 5 | Huynh-Ngoc and Fukai 1979 |
| coastal areas: | | | | | | |
| Hg-T | 19 | 2.25 | 1.4 - 5.6 | N-Tyrr. coast | 0 | Barghigiani et al., 1981 |
| Hg-Td | 24 | 6.3 | 1.4 - 8.0 | Tyrrh. coast | 0 | Ferrara et al., 1986 |
| Hg-T | 20 | 9.6 | 1.7 - 12.2 | Tuscan coast | 0 | Seritti et al., 1982 |
| Non Mediterranean | | | | | | |
| open sea: | | | | | | |
| Hg-T | 47 | 2.2 | +- 1.0 | N. Atlantic | 0 - 1730 | Olafson, 1983 |
| Hg-R | 81 | | 0.9 - 6.2 | North Sea | 0 | Baker, 1977 |
| Hg-R | 52 | 5 | 3.9 - 5.6 | Japan Sea | 0 - 1200 | Matsunaga et al., 1975 |
| coastal areas | | | | | | |
| Hg-T | ? | 7.9 | 3.4 - 22 | "UK seas" | 0 | Baker, 1977 |
| Hg-T | 4 | 5.1 | 3.2 - 7.4 | Suruga B.Jap | 0 | Fujita and Iwashima, 1981 |

Hg-T: total Hg
Hg-Td: total dissolved Hg (membrane filtered)
Hg-A: ASV, unfiltered at pH 2
Hg-R: reactive Hg (in acidified sample)
M: medium

Sediments: Not many data on open sea sediment concentrations have been collected in the Mediterranean Sea. When considering these data one has to bear in mind that the analytical procedures differ among authors and only few authors report on quality control. The use of different pretreatments (extraction methods) by the various authors makes the results not strictly comparable, but the order of magnitude can be assumed to be right. The few data available today show that 0.05 to 0.1 mg Hg-T $kg^{-1}$ DW may be considered a typical background value for the Mediterranean. Industrial sources and the frequent natural geochemical anomalies in the Mediterranean influence the Hg distribution in the marine sediments adjacent to these sources. Near river mouths, due either to anthropogenic or natural sources, sediments show higher levels. "Hot spots" (up to 5 ppm) have been observed near several coastal towns and most probably more "hot spots" will be found in a systematic survey. Near chlor-alkali plants and other petrochemical industries high concentrations (up to 200 ppm in the St. Gilla Lagoon, Cagliari, Sardinia) have been determined but the extention of these high Hg levels is limited in space. After 10 to 20 km background levels are reached again. High Hg levels are observed near geochemical anomalies and the high influence of rivers draining geochemical Hg anomalies such as the Mt. Amiata and the Idrija region (up to about 50 ppm) has been shown in the sediments adjacent to the river mouths (Tables 4 and 5) (For the understanding of table 5, see also fig. 2).

Biota: Comparing general data on the Hg-T concentration in biological species which present typical seafoods from the North Atlantic with those from the Mediterranean show that in general Mediterranean fishes have higher Hg levels (Tables 6 and 7).

In fact only the means of the Hg levels in plaice from the Atlantic are higher than 500 ug Hg-T $kg^{-1}$ FW, while the means of several of the Mediterranean species do exceed this level. Since Hg is an accumulative contaminant i.e. the Hg concentrations increase with age of specimen, better comparisons are obtained by confronting Hg concentrations in specimens of the same species caught in the Western Mediterranean and in the Atlantic. The first data, showing that mercury concentrations were higher in pelagic fishes from the Mediterranean than in the same species from the Atlantic, were published in the early seventies by Thibaud and Cumont and collaborators and were reviewed by Bernhard and Renzoni [6]. These data were later confirmed by data for different species which compare Hg concentrations versus weight of specimens. The clearest evidence comes from the Hg concentrations in bluefin tunas; it shows two distinct populations: a "high-mercury" and a "low-mercury" population [10]. The small tuna collected north of Sicily, the medium size tuna from the Adriatic and from the Ligurian Sea as well as a part of the large tuna caught in the tuna traps situated in Sicily and Sardinia belong to the "high-mercury" population. Another group of tuna belongs to the "low-mercury" population. It is important to note that tuna belonging to the "high-mercury" and the "low-mercury" populations were both caught off Sicily and off Sardinia, but in the Strait of Gibraltar only (!) tuna belonging to the "low-mercury" were caught. The migration pattern of bluefin tuna can explain the origin of these two tuna populations. Fisheries biologists studying these migration patterns have maintained for some time that Atlantic tuna enter the Mediterranean for spawning and leave again through the Strait of Gibraltar [7]. Thibaud [8] has analyzed several hundred tunas from the French Mediterranean coast and found that, with two exceptions, all belonged to the "high-Hg population". In order to understand better the differences in mercury levels between the Mediterranean and Atlantic biota Buffoni et al. [9] and Bernhard [10] constructed a very simple model of the mercury uptake and the loss of simplified tuna foodchain by using sardine data from the Western Mediterranean only. This model shows that methylmercury (MeHg) increases with the age (size) of the marine organism. This is best shown in the data on sardines [10].

Recently the prediction that the percentage of MeHg increases with age has been confirmed in muscle tissue of sardines, mackerels and bonitos [11]. The model also suggested that the higher concentrations in the Mediterranean fish should be due to higher Hg concentration in plankton and in sea water from the Mediterranean. Bernhard and Renzoni

TABLE 4
Selected mercury concentrations (mg kg $^{-1}$ DW) in "open-sea" sediments.
(Extract from MAP Technical Reports Series No. 28, table 16, Annex)

| depth | | | n | mean | range | | | location | reference |
|---|---|---|---|---|---|---|---|---|---|
| 2720 | | | 1 | 0.26 | | | | Alboran | Robertson et al., 1972 |
| ? | | | 51 | 0.23 | 0.01 | - | 0.64 | E. Gulf Lion | Arnoux et al., 1983 |
| ? | | | 43 | 0.11 | 0.01 | - | 0.27 | W. Gulf Lion | Arnoux et al., 1983 |
| ? | | | 14 | 0.38 | 0.07 | - | 0.23 | NW.Mediterranean | Arnoux et al., 1983 |
| ? | | | 17 | 0.13 | 0.16 | - | 0.57 | NW.Mediterranean | Arnoux et al., 1983 |
| 93 | - | 1715 | 9 | 0.1M | 0.05 | - | 0.24 | Tyrrhenian | Selli et al., 1973 |
| 390 | - | 3520 | 4 | 0.1M | 0.05 | - | 0.16 | Tyrrhenian | Selli et al., 1973 |
| 5 | - | 1195 | 20 | 0.1M | 0.07 | - | 0.97 | Adriatic | Selli et al., 1973 |
| 64 | + | 888 | 2 | | 0.05 | - | 0.1 | Adriatic | Selli et al., 1973 |
| 12 | - | 1200 | 38 | 0.05 | 0.01 | - | 0.16 | Adriatic | Kosta et al., 1978 |
| 2360 | | | 1 | 0.3 | | | | S.of Crete | Robertson et al., 1972 |

TABLE 5
Mercury concentrations in sediments of the Mediterranean
(Extract from MAP Technical Reports Series No. 28, table 17, Annex)

| | Region | Extraction method | Concentration ug g$^{-1}$ dry weight | Reference |
|---|---|---|---|---|
| I | Alboran Sea | Total | 0.26 (mean) | Robertson et al., 1972 |
| II | Ligurian coasts | $HNO_3$, HCl | 0.16-5.4 | Breder et al., 1981 |
| | Ebro delta | conc. $HNO_3$ | 0.065-1.1 | Obiols and Peiro, 1980 |
| | Area of Marseille | $HNO_3$ | 0.07-21 | Arnoux et al., 1980a, 1980b, 1980c |
| | Bay of Cannes | $HNO_3$, $HPO_4$ fraction 63 u | 0.1-0.4 | Ringot, 1982 |
| | Gulf of Nice | $HNO_3$, $HClO_4$ | 0.01-0.16 | Flatau et al., 1982 |
| | Catalan coasts | conc. $HNO_3$ | 0.2-1.0 | Peiro et al., 1980 |
| III | Santa Gilla lagoon, Cagliari | $HSO_4$, $HNO_3$ | 0.7-37 | Sarritzu et al., 1982 |
| IV | Tyrrhenian Sea | - | 0.05-0.24 | Selli et al., 1973 |
| | Tuscany Coast | - | | |
| | near Solvay plant | | 1.1-1.3 | Renzoni et al., 1973 |
| | 4 km S and N | | 0.1-0.8 | |
| | 10 km S and N | | 0.04-0.1 | |
| V | Gulf of Trieste | - | 1.4-14.8 | Majori et al., 1978 |
| | (close to cinnabar mine) | 19.4 | | |
| | Gulf of Venice | $H_2SO_4$ | 0.14-3.0 | Donazzolo et al., 1978 Angela et al., 1980 |
| | Kastela Bay Dalmatia (chlor-alkali plant) | Total | 8.5 | Stegnar et al., 1980 |
| | Adriatic Sea | Total | 0.07-0.97 | Robertson et al., 1972 |
| VIII | Evoikos Gulf Aegean Sea | 0.5 HCl fraction 55 u | 0.3-0.8 | Angelidis et al., 1980 |
| | Saronikos Gulf, Athens | Total | 0.5-1 | Grimanis et al., 1976 Papakostidis et al., 1975 |
| | Athens outfall | Total | 0.5-3 | |
| IX | Coasts of Turkey | $HNO_3$ | 0.019-0.48 | Tuncel et al., 1980 |
| X | Region of Alexandria (close to chlor-alkali plant) | conc. $HNO_3$ | 0.8 9-15 | Elsokkary, 1978 El Sayed & Halim, 1978 |
| | Haifa Bay | $HNO_3$ fraction 250 u | 0.008-073 | Krumgalz & Hornung, 1982 |
| | Hanigra to Hafifa | | 0.01-0.57 | Roth & Hornung, 1977 |

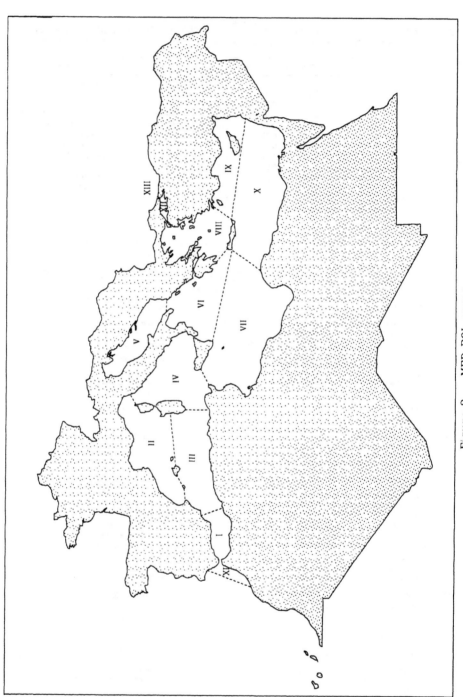

Figure 2. – MED POL areas

[6], Buffoni et al. [9] and Bernhard [10] suggested that the higher Hg levels in marine organisms observed in different Mediterranean regions are principally due to the inputs of mercury into the marine environment from the various Hg geochemical anomalies present in the Mediterranean area and that anthropogenic sources have significant but only very local influence. However, published data on sea water and plankton reveal no differences between samples collected in Mediterranean and Atlantic areas. This is primarily due to the fact that the analysis of Hg in sea water presents still great difficulties and hence the accuracy of the Hg determination is uncertain. These difficulties are evident from the large variability of the data published (Table 3).

TABLE 6
Mercury (ug kg$^{-1}$ FW) in some fish (muscle) and shellfish species (whole body).
(Extract from MAP Technical Reports Series No. 28, table 18, Annex)

| | Median of means and range of means | | | |
| --- | --- | --- | --- | --- |
| | median | range | location | references |
| plankton feeder | | | | |
| herring | 40 | 20-240 | N. Sea | ICES, 1974 |
| herring | 20 | 10-35 | N. Atl. | ICES, 1977a |
| herring | 40 | 10-23 | Irish coast | ICES, 1980 |
| "typical" | 40 | | | |
| sardine | 60 | 6-80 | N.Atl. | ICES, 1977a |
| sardine | 250 | 150-390 | Medit. | UNEP, 1980 |
| sprat | 65 | 60-140 | Irish c. | ICES, 1980 |
| capelin | 10 | 10-30 | N.Atl. | ICES, 1977a |
| anchovy | 160 | 145-180 | Medit. | UNEP, 1980 |
| feed on invertebrates | | | | |
| brown shrimp | 110 | 50-230 | N.Sea | ICES, 1974 |
| brown shrimp | 140 | 70-390 | N.Sea | ICES, 1977b |
| brown shrimp | 80 | 30-300 | N.Sea | ICES, 1977c |
| "typical" | 110 | | | |
| deep sea prawn | 25 | 20-30 | W.Greenl. | ICES, 1977a |
| Norway lobster | 960 | 290-970 | Medit. | UNEP, 1980 |
| cod | 100 | 30-480 | N.Sea | ICES, 1974 |
| cod | 100 | 60-300 | N.Sea | ICES, 1977a |
| cod | 40 | 40-50 | N.Atlantic | ICES, 1977a |
| cod | 260 | | Irish Sea | ICES, 1980 |
| cod | 140 | 70-370 | Irish Coast | ICES, 1980 |
| cod | 70 | 50-140 | NW.Atlantic | ICES, 1977a |
| cod | 80 | 70-90 | NW.Atlantic | ICES, 1980 |
| "typical" | 100 | | | |
| feed on crustaceans and fish | | | | |
| hake | 90 | 30-130 | N.Atlantic | ICES, 1977a |
| hake | | 30-850 | Mediterr. | UNEP, 1980 |
| haddock | 50 | 20-60 | Irish coast | ICES, 1980 |
| haddock | 50 | | NW.Atlantic | ICES, 1980 |
| whiting | 80 | 30-90 | Irish coast | ICES, 1980 |
| bluefin tuna | 715 | 300-1485 | Mediterr. | UNEP, 1980 |
| mullet | 190 | 55-1300 | Mediterr. | UNEP, 1980 |
| Greenl.halibut | 40 | 30-50 | N.Atlantic | ICES, 1977a |
| plaice | 90 | 20-260 | N.Sea | ICES, 1974 |
| plaice | 120 | 20-500 | N.Atlantic | ICES, 1977a |
| plaice | 25 | 10-80 | Irish coast | ICES, 1980 |
| "typical" | 90 | | | |
| sole | 150 | 50-320 | N.Atlantic | ICES, 1977a |

TABLE 7
Overall averages of mercury concentrations according to
UNEP sampling areas (fig. 2)
(Extract from MAP Technical Reports Series No. 28, table 19, Annex)

| Area | Species | n | ug kg$^{-1}$ FW mean | min | max |
|------|---------|---|------|-----|-----|
| II | Engraulis encrasicolus | 37 | 140 | 20 | 300 |
| | Mullus barbatus | 262 | 590 | 15 | 5600 |
| | M. surmuletus | 5 | 260 | 70 | 510 |
| | Mytilus galloprovincialis | 37 | 70 | 15 | 400 |
| | Nephrops norvegicus | 129 | 1080 | 350 | 3000 |
| | Sarda sarda | 14 | 1000 | 290 | 2300 |
| | Thunnus thynnus | 176 | 1100 | 20 | 6290 |
| | Xiphias gladius | 1 | 150 | | |
| III | M. surmuletus | 204 | 90 | 30 | 230 |
| | Perna perna | 192 | 76 | 20 | 370 |
| IV | E. encrasicolus | 44 | 157 | 65 | 380 |
| | M. barbatus | 185 | 1440 | 60 | 7050 |
| | M. galloprovincialis | 59 | 240 | 25 | 1260 |
| | N. norvegicus | 86 | 1110 | 60 | 2900 |
| | Thunnus alalunga | 8 | 215 | 90 | 336 |
| V | M. barbatus | 6 | 190 | 100 | 390 |
| | M. galloprovincialis | 26 | 870 | 25 | 7000 |
| VI | E. encrasicolus | 11 | 145 | 55 | 270 |
| | M. barbatus | 13 | 190 | 45 | 330 |
| | M. galloprovincialis | 12 | 75 | 35 | 145 |
| | N. norvegicus | 7 | 290 | 190 | 360 |
| | T. alalunga | 8 | 275 | 60 | 400 |
| VII | Lithophaga lithophaga | 5 | 165 | 80 | 290 |
| | M. barbatus | 11 | 165 | 30 | 280 |
| | Trachurus mediterraneus | 5 | 345 | 80 | 955 |
| VIII | Carcinus mediterraneus | 13 | 215 | 115 | 345 |
| | Merluccius merluccius | 10 | 315 | 60 | 840 |
| | Mugil auratus | 16 | 350 | 85 | 2500 |
| | Mugil cephalus | 3 | 165 | 70 | 300 |
| | M. barbatus | 127 | 175 | 15 | 1400 |
| | M. galloprovincialis | 175 | 105 | 5 | 920 |
| | Penaeus kerathurus | 10 | 175 | 75 | 475 |
| | T. thynnus | 7 | 370 | 70 | 890 |
| | T. mediterraneus | 3 | 340 | 320 | 365 |
| | X. gladius | 8 | 280 | 85 | 755 |
| IX | Boops salpa | 3 | 10 | 5 | 15 |
| | M. auratus | 39 | 170 | 1 | 5600 |
| | M. barbatus | 6 | 55 | 2 | 90 |
| | M. surmuletus | 13 | 35 | 1 | 80 |
| | M. galloprovincialis | 4 | 37 | 20 | 50 |
| | P. kerathurus | 7 | 20 | 10 | 50 |
| | Upenaeus moluccensis | 7 | 200 | 110 | 430 |
| | Boops boops | 5 | 135 | 40 | 430 |
| | Dentex dentex | 6 | 385 | 220 | 480 |
| | D. gibbosus | 12 | 140 | 100 | 180 |
| | Donax trunculus | 42 | 210 | 35 | 910 |
| | Epinephelus aenus | 4 | 250 | 100 | 400 |
| | Merluccius merluccius | 6 | 150 | 31 | 260 |
| | M. barbatus | 168 | 140 | 30 | 475 |
| | Pagellus acarne | 7 | 190 | 70 | 340 |
| | P. erythrinus | 112 | 205 | 55 | 805 |

TABLE 7 (Cont'd./..2)
Overall averages of mercury concentrations according to
UNEP sampling areas (fig. 2)
(Extract from MAP Technical Reports Series No. 28, table 19, Annex)

| Area | Species | n | ug kg$^{-1}$ FW mean | min | max |
|------|---------|---|------|-----|-----|
| X | Surida undosquamis | 143 | 135 | 40 | 650 |
|   | Sphyraena sphyraena | 7 | 165 | 80 | 245 |
|   | T. mediterraneus | 48 | 95 | 10 | 415 |
|   | U. moluccensis | 120 | 440 | 40 | 1120 |
| XI | M. surmuletus | 5 | 150 | 15 | 380 |
|    | M. galloprovincialis | 3 | 190 | 20 | 290 |
|    | T. thynnus | 1 | 550 | | |
| XII | M. merluccius | 3 | 815 | 780 | 850 |
|     | M. barbatus | 3 | 215 | 210 | 230 |
|     | M. galloprovincialis | 3 | 160 | 140 | 170 |
|     | P. erythrinus | 3 | 220 | 210 | 225 |
|     | Parapenaeus longirostris | 3 | 300 | 270 | 350 |
|     | T. mediterraneus | 3 | 345 | 340 | 350 |

The highest Hg concentrations have been observed in marine mammals
(Table 8). A comparison between animals of the same species and of about
the same size shows that here also Mediterranean specimens have higher Hg
levels than Atlantic specimens.
Man: Astier-Dumas and Cumont [12] studied the seafood consumption
in four French regions. They found that persons eating more than three
meals week$^{-1}$ had higher Hg levels in their hair (mean (5) = 7.60 ± 3.4
ppm ) than persons consuming less seafood (mean (6) = 1.1 ± 0.6 ppm).
Paccagnella et al. [13] selected the population of Carloforte
(Sardinia) for an epidemiological study, because its average consumption
of seafood was about 4 times the national Italian average and because,
during the summer months, fresh tuna meat was consumed. From 6,200
residents 195 persons chosen at random agreed to give information about
their food habits, take a medical examination and allow a blood and hair
analysis. Based on mercury analysis of tuna and other seafoods and the
seasonal seafood consumption patterns it was estimated that the average
weekly intake of mercury in the summer was 150 ug and in the winter 100
ug. The group with the highest consumption (14 seafood meals per week)
had an estimated weekly mercury intake of 700 ug in the summer and 460 ug
in the winter. Their average hair mercury level was 11 mg Hg kg$^{-1}$
(range: from "not detected" to 60 mg kg$^{-1}$), which fits well with the
estimate that of an average intake of 300 ug Hg week$^{-1}$ the hair level
would be about 6 mg Hg kg$^{-1}$ [14].
Riolfatti [15] compared hair mercury levels in an inland town with
a coastal town, where 13% of the 52 persons examined had consumed more
than four fish meals per week. One man in the coastal town had hair
levels which fell within the range of possible earliest effects [14] i.e.
his hair concentration was about 45 mg Hg kg$^{-1}$ and six others reached hair
concentrations between 16 and 20 mg kg$^{-1}$. In the inland town relatively
high hair concentrations were also observed. One woman had about 30 mg
kg$^{-1}$ and three had levels between 16 and 25 mg kg$^{-1}$ despite the fact that
none of the persons examined in the inland town had consumed more than 2
fish meals per week. As no quality assurance of the analytical method
was carried out and the results are unusually high no conclusions can be
drawn.
Bacci et al. [16] studied the total and methylmercury
concentrations in blood, urine, hair and nails of 16 persons from the
town of Vada, who consumed from 0 to more than 6 meals of seafood per
week. The fish came from the banks of the Vada river about 10 km West of
the then operating Solvay chlor-alkali plant. As expected, the mercury
concentrations increased with the amount of seafood meals consumed. The
concentration in the hair ranged from 4 to 110 mg Hg kg$^{-1}$.

TABLE 8
Mercury concentrations in pelagic mammals from the Mediterranean
and Atlantic
(Extract from MAP Technical Reports Series No. 28, table 22, Annex)

| Species | sex | age | size cm | –ug muscle | kg-T fat | kg$^{-1}$ FW – liver | sample location and date |
|---|---|---|---|---|---|---|---|
| **Atlantic:** | | | | | | | |
| Phocoena phocoena | M | adult | 172 | 6750 | 770 | 61000 | La Rochelle (May 1972) |
| Delphinus delphis | F | young | 125 | 890 | 710 | 900 | Ile de Ré (July 1972) |
| | F | adult | 140 | 600 | 20 | 980 | Pyrénées Atl. (July 1973) |
| | F | adult | 165 | 910 | 27 | 1430 | Pyrénées Atl. (April 1973) |
| | M | adult | 185 | 1840 | 220 | 220 | Landes (July 1973) |
| | F | adult | 210 | 6250 | 2650 | 4850 | Gironde (May 1972) |
| | M | 15 y | 220 | 2180 | 2780 | 66700 | Tropic Atl. 1975 |
| **Mediterranean:** | | | | | | | |
| D. delphis | M | 12 y | 205 | 1450 | 3900 | 604000 | Mediterranean 1973 |
| Stenella coeruleo. | F | adult | 168 | 1950 | 1800 | 39850 | Iles d'Hyères (Feb. 1973) |
| | M | adult | 210 | 23800 | 6000 | 344900 | Lavandou (Var) (April 1973) |
| Grapus priseus | F | adult | 300 | 16000 | 1700 | 905000 | Cacalastre (Var) (July 1973) |
| Tursiops truncata | ? | | 140* | 41000 | - | - | Pescara (1971) |
| | M | 6-18m | 160 | 2200 | 310 | 14600 | Mediterranean (1973) |
| | M | 25 y | 330 | 24000 | 4400 | 293000 | Mediterranean (1973) |
| **Atlantic:** | | | | | | | |
| Globicephala melaena | F | young | 300 | 640 | 50 | 900 | Gironde (April 1972) |
| | M | adult | 490 | 5300 | 860 | 860 | Charente (August 1972) |
| **Mediterranean:** | | | | | | | |
| G. melaena | F | adult | 390 | 13100 | 1290 | 670000 | Cros de Cagne (Alp. Mar.) (July 1973) |
| Physeter catodan | M | ? | 800 | 4050 | 3150 | - | Bonifacio (Cors. (Dec.1972) |

*)   size in kg
M = male; F = female; y = year; m = month

Nauen et al. [17] estimated the amount of mercury intake from a food consumption study on the basis of a survey in three Italian locations. Information on individual seafood consumption over a period of 20 days was matched with analytical data on Hg levels in the fish and shellfish consumed. Special attention was given to fishermen and their families. Applying a consumer risk simulation model the authors found that a high percentage of the persons interviewed exceeded their daily allowance, among them many children. In fact the maximum intake for an individual was estimated for a 3-year old child which reached 8.6 times the daily tolerable allowance.

## Other Trace Elements

Sea water: Trace metals determined in sea water samples from the Mediterranean showed that Cd ranges from 0.02 to 0.7, Cu from 0.04 to 5.8 ug l$^{-1}$ and Zn from 0.02 to 10 ug l$^{-1}$ [18]. Kremling and Petersen [19] found narrower ranges: Cu from 0.11 to 0.43 ug l$^{-1}$, Fe from 0.12 to 0.46 ug l$^{-1}$, Mn from 0.05 to 0.84 ug l$^{-1}$ and Zn from 0.2 to 0.76 ug l$^{-1}$, but for Cd much higher levels from 3.2 to 30.5 ug l$^{-1}$. In coastal waters, concentrations vary widely with Pb levels up to 20 ug l$^{-1}$, Cd up to 1.4 ug l$^{-1}$, Cu up to 50 ug l$^{-1}$, Cr up to 6.7 ug l$^{-1}$, Ni up to 9 ug l$^{-1}$ and Zn up to 36 ug l$^{-1}$ in very polluted areas [20] and [21] (Tables 9 and 10). Recent preliminary data seem to show a difference in Cd and Pb concentrations in waters from the Mediterranean and from the Atlantic in the Strait of Gibraltar and in waters of the western and eastern basin in the Strait of Sicily [22]. Sea water concentrations for arsenic range from about 1 to 4 ug As-T l$^{-1}$. "Reactive As" ranges between 1 and 3.5 ug As l-1 [23]. In other oceans As-T levels range from about 1 to 2 ug l$^{-1}$ [24]. Due to the lack of sea water standards for trace elements at the very low sea water concentrations, the accuracy of the sea water data is uncertain.

TABLE 9
Cadmium, Copper, Lead and Zinc concentrations in open waters of the Mediterranean (ug l$^{-1}$)
(Extract from MAP Technical Reports Series No. 28, table 2, Annex)

| REGION | METHOD | Cadmium | Copper | Lead | Zinc | REFERENCE |
|--------|--------|---------|--------|------|------|-----------|
| I-II | | 0.004 | 0.11 | - | - | Boyle et al., 1984 |
| II | ASV | 0.15 | 0.4 | - | 2.7 | Huynh-Ngoc and Fukai, 1978 |
| | Dowex/ Extraction/ AAS | 0.06 | 1.6 | - | - | Frache et al., 1980 |
| II | | 0.008 | - | 0.05-0.14 | - | Copin-Montegut et al., 1984 |
| III | ASV | 0.005-0.010 | 0.06-0.13 | 0.025-0.075 | - | Laumond et al., 1982 |
| | ASV | 0.11 | 0.10 | - | 1.2 | Huynh-Ngoc and Fukai, 1978 |
| IV | ASV | 0.11 | 0.18 | - | 0.9 | " |
| | ASV | 0.05-0.09 | 0.13-0.19 | 0.018-0.09 | - | Nürnberg, 1977 |
| IV-VI-VII | | 0.010 | 0.15 | - | - | " |
| VI-VII | ASV | 0.15 | 0.7 | - | 1.8 | Huynh-Ngoc and Fukai, 1978 |
| VIII | ASV | 0.07 | 0.3 | - | 3 | " |
| X | ASV | 0.04 | 0.04 | - | 0.9 | " |

TABLE 10
Chromium and Nickel concentrations in Mediterranean coastal waters (ug l$^{-1}$)
(Extract from MAP Technical Reports Series No. 28, table 3, Annex)

| REGION | METHOD Chemical form | Chromium | Nickel | REFERENCE |
|--------|----------------------|----------|--------|-----------|
| II Ligurian coasts | Dowex A-1/AAS Filtration/Dowex A-1/AAS | | 0.15-2.9 | Frache et al., 1976 |
| | Dissolved Particulate | | 0.27-9.0 0.15-0.89 | Baffi et al., 1982 |
| V VIII | X-ray emission Filtration/Chelex 100/AAS | 0.68 | 1.3 | Marijanovic et al., 1982 |
| Evoikos, Gera Gulfs, Greece Northern Greece | Dissolved Particulate APDC-MIBK Extraction/AAS | 6.2-6.7 1.5-2.5 | 1.9 0.5 0.5-1.5 | Scoullos and Dasenakis, 1982 Scoullos et al., 1982 Fytianos and Vasilikiotis, 1982 |

Sediments: Sediment concentrations are several orders of magnitude higher than sea water concentrations. Data on trace element levels in sediments from open sea areas are rare. Whitehead et al. [25] examined coastal sediments taken during the Calypso cruise in 1977 and data from the literature. These authors suggest the following levels (in mg kg$^{-1}$ DW) as background: Cr - 15, Cu - 15, Zn - 50, Cd - 0.15, Pb - 25. Concentrations for arsenic in coastal sediments can reach up to 10 ppm

(DW) which compares well with concentrations observed in other coastal sediments for which a range of 3 to 14 ppm (DW) are reported (review: WHO, 1986). Deep sea sediments are reported to have higher levels of arsenic: up to 120 ppm (DW) with a world average of 20 to 30 ppm. The data available show that higher levels are found in estuarine sediments near discharges of industry and that the levels decrease towards the open sea. In several cases multi-element analysis has been carried out. However, it is not just anthropogenic sources that can influence sediment concentrations. The mineralogical composition of the marine sediments and the natural geochemical composition of adjacent coastal terrestrial environments can increase trace element levels in coastal sediments above background levels [21] and [26]. In Tables 11 and 12 the levels for several coastal areas are summarized.

**Biota**: Following the "mussel-watch" programme, the mussel M. galloprovincialis has been used as a pollution indicator in the MED POL phase I programme. A wide range of concentrations has been observed. Background concentrations for mussels are difficult to establish because they can reflect very local conditions. Cu concentrations in mussels of four MED POL areas (II, IV, V, VIII) (Fig. 2) have a mean of all samples of about 1300 ug kg$^{-1}$ FW with ranges comparable to those reported for the North Sea. The mean of all Cd concentrations in mussels was 120 ug kg$^{-1}$ FW (excluding 5% of the highest values). The mean of all Zn concentrations in mussels was 27000 ug kg$^{-1}$ FW. Also here, the ranges are similar to those observed for the North Sea and the mean of all Pb levels was 800 ug kg$^{-1}$ FW [21], [27], [28] and [29]. Cd, Cu, Zn and Pb levels in M. barbatus, another pollution indicator of the MED POL programmes, are considerably lower for Cd (mean of all samples in areas II, IV, VII, VIII, X: 46 ug kg$^{-1}$ FW), Cu (mean of all samples in areas II, IV, VII, VIII, X: 600 ug kg$^{-1}$ FW), Zn (mean of all samples in areas II, IV, VII, VIII, IX, X: 3900 ug kg$^{-1}$ FW) and Pb (mean of all samples in areas II, X: 70 ug kg$^{-1}$ FW). Several other organisms have been analyzed; their data together with details of the above mentioned data are shown in Tables 13 to 21. Total arsenic (AS-T) concentrations in marine fishes from the Mediterranean range between 8 and 70 mg As-T kg$^{-1}$ FW with Sepia officinalis reaching up to 370 mg As-T kg$^{-1}$ FW [23]. The same authors observed levels up to 630 mg As-T in the flatfish Limanda limanda in the North Sea. "Typical levels" for fish should be in the range of 1 to 10 mg As-T kg$^{-1}$ FW [24].

## Halogenated hydrocarbons

**Sea water**: Sea water samples had a concentration range from 0.2 to 38 ng PCBs l$^{-1}$ (Table 22). Risebrough et al. [30] found considerably lower PCB levels in sea water samples near the French coast showing the difficulties inherent in comparing data from different authors without intercalibration. In the Northern Adriatic coastal waters in 50 samples analyzed between 1977 and 1978 most concentrations were below the detection limits of 0.05 ng DDT l$^{-1}$ and 0.1 ng PCBs l$^{-1}$ [31]. Lindane levels off-shore in the eastern basin ranged from 0.06 to 0.12 ng l$^{-1}$. The higher concentrations were observed near terrestrial run-offs and river inputs. Particulate matter had higher concentrations than the sea water dissolved phase.

**Sediments**: Concentrations of PCBs in the top centimeter of open sea cores sampled in the Eastern and Western Mediterranean ranged from 0.8 to 9.0 ug PCBs kg$^{-1}$DW (Table 22). So far in coastal sediments only examples for "hot spots" can be sited. In many other similar situations high concentrations are likely to be found. The examples concern high sediment concentrations near the sewage outfalls of Marseille (up to 16000 ug PCBs kg$^{-1}$DW) and Athens (up to 800 ug PCBs kg$^{-1}$DW), near larger towns such as Nice (up to 1165 ug PCBs kg$^{-1}$ DW), Naples (up to 3200 ug PCBs kg$^{-1}$ DW) and Augusta (up to 460 ug PCBs kg$^{-1}$DW). However, in the case of the Marseille outfall, these high levels drop to background levels at about 10 km from the source [32]. Coastal sediment

TABLE 11

Cadmium, Copper, Lead and Zinc concentrations in Mediterranean sediments ($\mu g\ g^{-1}$ dry weight)
(Extract from MAP Technical Reports Series No. 28, table 4, Annex)

| REGION | METHOD | Cadmium | Copper | Lead | Zinc | REFERENCE |
|---|---|---|---|---|---|---|
| II | | | | | | |
| Var lagoon, France | HF-HClO4-HNO3 | 3.7 | 15.4 | 26.4 | - | Chabert and Vicente, 1980 |
| Coastal lagoon, Spain | 63um | 10-32 | 10-94 | 200-2000 | 500-6200 | De Leon et al., 1982 |
| Spain | Conc.HNO3 | 0.1-0.3 | - | 10-50 | - | Peiro et al., 1982 |
| River Ebro Delta | HNO3 | 0.12-0.37 | 7.9-21.5 | 22-48 | 33-104 | Obiols and Peiro, 1980 |
| River Rhône Delta | HNO3-HClO4 | 0.25-5 | 20-55 | 9 | 90-140 | Added et al., 1980; Cauwet and Monaco, 1982; Badie et al., 1982 |
| | | | | | | |
| Marseille | 200um HCl-HNO3 | 1.8-3 | 29-34 | 28-1250 | 120-2550 | Arnoux et al., 1980a |
| Cannes | 63um HNO3-H3PO4-HCl | 1.8-7 | 15-80 | 30-100 | 50-300 | Ringot, 1982 |
| Gulf of Nice | HNO3-HCl | 0.7-2.4 | 2.1-32 | 3-112 | - | Flateu et al., 1982 |
| Italian Estuaries | HNO3-HCl | 0.21-0.55 | 33-53 | 30-43 | - | Breder, 1980 |
| Offshore sediments | HNO3-HCl | | 30-49 | 10-28 | 130-260 | Arnoux et al., 1980c,1982 |
| | HNO3 | 0.7-1.7 | | | | Frignani and Giordani, 1983 |
| III | | | | | | |
| Coast of Spain | - | 0.02-10 | 4-230 | 23-3300 | 27-1050 | De Leon et al., 1984 |
| IV | | | | | | |
| Cagliari lagoon | 0.4NHCl | 0.5-2.5 | 10-70 | 64-670 | - | Contu et al., 1984 |
| Offshore sediments | HNO3 | | 10-44 | 19-94 | 20-56 | Frignani and Giordani, 1983 |
| V | | | | | | |
| Gulf of Trieste | ? | 0.3-5.3 | 9-139 | 18-470 | 27-650 | Majori et al., 1978 |
| Gulf of Venice | HNO3 | 0.1-3.1 | 34-37 | 5-54 | 48-450 | Angela et al., 1980 |
| Kastela Bay Yugoslavia | NAA 100um | | 14-42 | - | 53-1300 | Stegnar et al., 1980 |
| River Po Delta | HNO3 | 0.16-1.7 | 1.3-50 | 9.-73 | 24-244 | Facardi et al., 1984 |
| Mali Ston, Yug. | NAA | 0.08-0.22 | 13-22 | - | 40-100 | Vukadin et al., 1984 |
| Offshore sediments | HNO3 | 0.8-1.2 | 15-30 | 21-43 | 54-78 | Frignani and Giordani, 1983 |
| VI | | | | | | |
| Patraikos Gulf, Greece | HF-HNO3 HClO4 | - | 23-100 | 10-40 | 280-430 | Varnavas and Ferentinos, 1982 |
| Kalamata Bay,Greece | HF-HNO3-HClO4 | - | 11-56 | 8-40 | - | Varnavas et al., 1984 |
| Gulf of Catania | HNO3 | 2.2-4.6 | 3.8-2.5 | 4.5-19 | 25-236 | Castagna et al., 1983 |
| Offshore sediments | HNO3 | 0.6-1.1 | 24-29 | 22.27 | 55-78 | Frignani and Giordani, 1983 |

TABLE 11 (Cont'd./..2)

Cadmium, Copper, Lead and Zinc concentrations in Mediterranean sediments (ug g$^{-1}$ dry weight)
(Extract from MAP Technical Reports Series No. 28, table 4, Annex)

| REGION | METHOD | Cadmium | Copper | Lead | Zinc | REFERENCE |
|---|---|---|---|---|---|---|
| VIII | | | | | | |
| Thermaikos-Kavala Gulf, Greece | 63 um HNO$_3$-HClO$_4$ | 0.6-1.1 | 0.6-2.3 | 6-28 | 10-28 | Fytianos & Vasilikiotis, 1982 |
| Evoikos Gulf, Greece | 61 um 0.5N HCl | - | 9 | 37 | 20 | Scoullos and Dasenakis, 1982 |
| Gera Gulf, Greece | 55 um 0.5N HCl | - | - | - | 7-95 | Angelidis et al., 1980 |
| Saronikos Gulf, Greece | 61 um 0.5N HCl | - | 8-160 | 9-122 | 12-390 | Scoullos et al., 1982 |
| Thermaikos Gulf, Greece | 55 um 0.5N HCl | - | - | - | 5-1360 | Angelidis et al., 1982 |
| Thermaikos Gulf, Greece | 45 um HNO$_3$ | 0.40-2.5 | 10-50 | 25-130 | 8-240 | Voutsinou-Taliadouri, 1982 |
| Pagassitikos Gulf, Greece | 45 um HNO$_3$ | - | 30 | 30 | 130 | Voutsinou-Taliadouri, 1982 |
| East Aegean Offshore | 45 um HNO$_3$ | - | 20 | 15 | 40 | Voutsinou-Taliadouri, 1982 |
| Izmir Bay | HCl-HNO$_3$ | 0.2-49 | 14-870 | 20-280 | 53-860 | Uysal and Tuncer, 1984 |
| IX | | | | | | |
| Erdemli, Turkey | HNO$_3$-HClO$_4$-HF | - | 31 | 57 | 65 | Balkas et al, 1978 |
| Alexandria | ? | 2.8 | 48 | 190 | 180 | El Sokkary, 1978 |
| Alexandria Harbour | HNO$_3$-HCl | 2 | 27 | - | 53 | El Sayed et al., 1980 |
| Abu Kir Bay, Egypt | HNO$_3$ | 2 | 12 | - | 100 | Saad et al., 1980 |
| River Nile delta | HF-HNO$_3$ | - | 5-77 | - | 2-120 | Moussa, 1982 |
| | HCl-1N | - | 6-74 | - | 20-100 | Tomma et al., 1980 |
| Cilician basin | HF-NHO$_3$-HClO$_4$ | - | 33-50 | - | 54-81 | Ozkan et al., 1980 |
| X | | | | | | |
| Damietta estuary Egypt | HNO$_3$ | 0.16-2 | 29-280 | - | 20-425 | Saad and Fahmy, 1984 |
| Western Harbor Alexandria | HNO$_3$-HClO$_4$ | 7-64 | 30-1890 | - | 23-470 | Saad et al., 1984 |
| XIII | | | | | | |
| Black Sea, Nearshore | HNO$_3$ | 1.3-4.8 | 10-100 | 22-88 | 37-250 | Pecheanu, 1982 |
| Offshore | | 2.8 | 52 | 37 | 75 | Pecheanu, 1982 |

## TABLE 12
Chromium, Manganese and Nickel concentrations in Mediterranean sediments (ug g$^{-1}$ dry weight)
(Extract from MAP Technical Reports Series No. 28, table 5, Annex)

| REGION | METHOD | Chromium | Manganese | Nickel | REFERENCE |
|---|---|---|---|---|---|
| **II** | | | | | |
| River Ebro Delta | $HNO_3$ | 9.5-20 | 180-630 | 22-47 | Obiols and Peiro, 1980 |
| Rhône Delta | $HNO_3$-$HCLO_4$ | 30-50 | - | 20-28 | Added et al., 1980 |
| Marseille | " | 7-230 | - | - | Arnoux et al., 1980a, 1980b |
| Gulf of Fos | 200 um, $HNO_3$-HCl | 27-32 | - | - | Arnoux et al., 1980a, 1980b |
| Offshore sediments | $HNO_3$-HCl | 35-65 | 700-1300 | - | Arnoux et al., 1980c, 1982 |
| Offshore sediments | $HNO_3$ | 28-37 | 280-1500 | - | Frignani and Giordani, 1983 |
| **IV** | | | | | |
| Offshore sediments | $HNO_3$ | 9-26 | 260-2560 | 9-46 | Frignani and Giordani, 1983 |
| **V** | | | | | |
| Gulf of Trieste | ? | 67-75 | 304-593 | 8-68 | Majori et al., 1978 |
| Gulf of Venice | $HNO_3$ | 10-64 | - | 5-40 | Angela et al., 1980 |
| Offshore sediments | $HNO_3$ | 31-60 | 52-2340 | 46-122 | Frignani and Giordani, 1983 |
| **VI** | | | | | |
| Patraikos Gulf, Greece | HF-$NHO_3$-$HClO_4$ | 70-210 | 83-1700 | 69-168 | Varnavas and Ferentinos, 1982 |
| Offshore sediments | $HNO_3$ | 31-33 | | 48-53 | Frignani and Giordani, 1983 |
| **VIII** | | | | | |
| Evoikos Gulf, Greece | 61 um, 0.5N HCl | 87 | 308 | 60 | Scoullos and Dasenakis, 1982 |
| Gera Gulf, Greece | 61 um, 0.5N HCl | 7.9-1830 | 80-570 | - | Scoullos et al., 1982 |
| Saronikos Gulf, Greece | 55 um, 0.5N HCl | 45-480 | - | - | Angelidis et al., 1982 |
| Thermaikos Gulf, Greece | 45 um, $HNO_3$ | 90-235 | - | 100-335 | Voutsinou-Taliadouri, 1982 |
| Pagassitikos Gulf, Greece | " | 110 | - | 165 | " |
| East Aegean, Offshore | " | 85 | - | 145 | " |
| **IX** | | | | | |
| Erdemli, Turkey | HF-$HNO_3$-$HClO_4$ | 534-595 | 115-787 | 79-586 | Ozkan et al., 1980 |
| **X** | | | | | |
| River Nile Delta | HF-$HNO_3$ | 12-150 | - | 10-100 | Moussa, 1982 |

TABLE 13
Copper concentrations (ug kg$^{-1}$ FW) in Mytilus galloprovincialis ug kg$^{-1}$
and Mullus barbatus from various MED POL areas
(Extract from MAP Technical Reports Series No. 28, table 6, Annex)

| Region | No. of samples | Minimum | Maximum | Mean | Standard Deviation |
|---|---|---|---|---|---|
| M. galloprovincialis: | | | | | |
| II | 55 | 504 | 4800 | 1500 | 900 |
| IV | 85 | 70 | 6000 | 1900 | 1100 |
| V | 58 | 163 | 4400 | 1000 | 900 |
| VIII | 13 | 750 | 2800 | 1600 | 600 |
| | | | | | |
| M. barbatus: | | | | | |
| II | 153 | 200 | 1300 | 405 | 172 |
| IV | 208 | 2.5 | 1000 | 380 | 127 |
| VII | 10 | 360 | 2700 | 930 | 684 |
| VIII | 60 | 220 | 1470 | 600 | 280 |
| X | 23 | 69 | 2550 | 800 | 560 |

TABLE 14
Average copper concentrations (ug kg$^{-1}$ FW) and standard deviation
in Mediterranean marine organisms from various MED POL areas
(Extract from MAP Technical Reports Series No. 28, table 7, Annex)

| Species | No. of samples | Mean | Standard Deviation |
|---|---|---|---|
| Mytilus galloprovincialis | 204 | 1300 | 700 |
| Donax trunculus | 19 | 3500 | 1800 |
| Nephrops norvegicus | 303 | 5700 | 1900 |
| Parapenaeus longirostris | 22 | 8500 | 8000 |
| Penaeus kerathurus | 12 | 5200 | 2700 |
| Engraulis encrasicolus | 97 | 990 | 560 |
| Mugil auratus | 31 | 700 | 960 |
| Mullus barbatus | 444 | 400 | 140 |
| Mullus surmuletus | 20 | 600 | 540 |
| Sarda sarda | 27 | 2100 | 1700 |

TABLE 15
Zinc concentrations (ug kg$^{-1}$ FW) in Mytilus galloprovincialis
and Mullus barbatus from various MED POL areas
(Extract from MAP Technical Reports Series No. 28, table 8, Annex)

| Region | No. of samples | Minimum | Maximum | Mean | Standard Deviation |
|---|---|---|---|---|---|
| M. galloprovincialis: | | | | | |
| II | 26 | 13000 | 60200 | 28000 | 10700 |
| IV | 84 | 3150 | 63000 | 34000 | 11200 |
| V | 58 | 2500 | 64250 | 17000 | 12000 |
| VIII | 21 | 9200 | 97700 | 45000 | 24600 |
| | | | | | |
| M. barbatus: | | | | | |
| II | 132 | 100 | 7100 | 4000 | 970 |
| IV | 221 | 400 | 7000 | 4000 | 1000 |
| VII | 11 | 2700 | 5800 | 4300 | 860 |
| VIII | 40 | 2570 | 6890 | 3500 | 800 |
| IX | 12 | 3660 | 7400 | 5100 | 1040 |
| X | 23 | 3060 | 5870 | 4400 | 650 |

TABLE 16
Average zinc concentrations and standard deviation (ug kg$^{-1}$ FW) in
Mediterranean marine organisms from all MED POL areas
(Extract from MAP Technical Reports Series No. 28, table 9, Annex)

| Species | No. of samples | Mean | Standard Deviation |
|---|---|---|---|
| Mytilus galloprovincialis | 179 | 27000 | 13000 |
| Donax trunculus | 17 | 21000 | 17000 |
| Nephrops norvegicus | 279 | 15000 | 2800 |
| Parapenaeus longirostris | 19 | 11000 | 3400 |
| Penaeus kerathurus | 22 | 22000 | 16000 |
| Carcinus mediterraneus | 13 | 41000 | 29000 |
| Engraulis encrasicolus | 75 | 18000 | 6700 |
| Mugil auratus | 66 | 10600 | 15000 |
| Mullus barbatus | 435 | 3900 | 900 |
| Mullus surmuletus | 24 | 10000 | 14000 |
| Upeneus molluccensis | 13 | 2900 | 1100 |

TABLE 17
Lead concentrations (ug kg$^{-1}$ FW) in Mytilus galloprovincialis
and in Mullus barbatus from various MED POL areas
(Extract from MAP Technical Reports Series No. 28, table 10, Annex)

| Region | No. of samples | Minimum | Maximum | Mean | Standard Deviation |
|---|---|---|---|---|---|
| M. galloprovincialis: | | | | | |
| II | 101 | 50 | 6800 | 600 | 790 |
| IV | 85 | 50 | 16100 | 1800 | 2400 |
| V | 92 | 50 | 7825 | 840 | 1300 |
| VIII | 80 | 55 | 8260 | 1100 | 1500 |
| M. barbatus: | | | | | |
| II | 173 | 23 | 243 | 60 | 31 |
| X | 22 | 145 | 610 | 370 | 121 |

TABLE 18
Average lead concentrations and standard deviation (ug kg$^{-1}$ FW)
in Mediterranean marine organisms from all MED POL areas
(Extract from MAP Technical Reports Series No. 28, table 11, Annex)

| Species | No. of samples | Mean | Standard Deviation |
|---|---|---|---|
| Mytilus galloprovincialis | 344 | 800 | 800 |
| Donax trunculus | 19 | 1200 | 650 |
| Mullus barbatus | 435 | 70 | 45 |
| Thunnus thynnus | 53 | 117 | 170 |

TABLE 19
Cadmium concentrations (ug kg$^{-1}$ FW) in <u>Mytilus galloprovincialis</u>
and <u>Mullus barbatus</u> in various MED POL areas
(Extract from MAP Technical Reports Series No. 28, table 12, Annex)

| Region | No.of samples | Minimum | Maximum | Mean | Standard Deviation |
|---|---|---|---|---|---|
| <u>M. galloprovincialis</u> | | | | | |
| II | 105 | 40 | 1060 | 190 | 120 |
| V | 72 | 25 | 475 | 160 | 100 |
| VI | 25 | 24 | 52 | 38 | 6 |
| VIII | 76 | 5 | 780 | 100 | 124 |
| <u>M. barbatus</u>: | | | | | |
| II | 136 | 1.0 | 590 | 50 | 90 |
| VI | 50 | 5.0 | 52 | 26 | 14 |
| VII | 11 | 5.5 | 49 | 17 | 15 |
| VIII | 46 | 15 | 162 | 47 | 39 |
| X | 21 | 14 | 65 | 39 | 14 |

TABLE 20
Average cadmium concentrations and standaard deviation (ug kg$^{-1}$ FW)
in Mediterranean marine organisms
(Extract from MAP Technical Reports Series No. 28, table 13, Annex)

| Species | no. of samples | Mean | Standard Deviation |
|---|---|---|---|
| <u>Mytilus galloprovincialis</u> | 265 | 120 | 83 |
| <u>Mytilus edulis</u> | 10 | 85 | 34 |
| <u>Donax trunculus</u> | 16 | 80 | 26 |
| <u>Nephrops norvegicus</u> | 61 | 50 | 39 |
| <u>Parapenaeus longirostris</u> | 27 | 46 | 55 |
| <u>Engraulis encrasicolus</u> | 81 | 34 | 25 |
| <u>Merluccius merluccius</u> | 27 | 63 | 34 |
| <u>Mugil auratus</u> | 10 | 47 | 85 |
| <u>Mullus barbatus</u> | 318 | 34 | 28 |
| <u>Mullus surmuletus</u> | 218 | 140 | 83 |
| <u>Thunnus alalunga</u> | 38 | 23 | 6.5 |
| <u>Thunnus thynnus</u> | 111 | 38 | 43 |

concentrations from the Central Mediterranean had the following levels:

| | DDT total mean | range | HCH total mean | range |
|---|---|---|---|---|
| area IV | 7.8 | (0.6 - 26.9) | 2.5 | (0.1 - 27.4) |
| V | 4.5 | (NDc - 47.8) | 1.1 | (0.2 - 4.6) |
| V | 11.5 | (NDc - 39.4) | 0.9 | (0.5 - 1.1) |
| VI | 12 | (2.4 - 35.5) | 0.1 | (0.1 - 2.6) |
| VIII | 245 | (7.1-1895 ) | 0.6 | (0.4 - 0.8) |
| IX | 12 | (0.4 - 29 ) | 0.2 | (0.2 - 0.3) |
| X | 390 | (1 -780 ) | 0.7 | |

## TABLE 21

Heavy metals (ug g$^{-1}$ dry weight*) in <u>Mytilus</u> from different regions of the mediterranean Sea. Values given are ranges

(Extract from MAP Technical Reports Series No. 28, table 14, Annex)

| region | Cadmium | Copper | Zinc | Lead | Nickel | Chromium | Silver | Iron | Mercury | Refs see |
|---|---|---|---|---|---|---|---|---|---|---|
| North-west Mediterranean (Ligurian Sea) | 0.4-5.9 | 2.4-154 | 97-644 | 2.4-117 | 0.9-14.1 | 0.5-28.8 | 0.1-18.9 | 149-2200 | 0.18-0.96 | 1, 2 |
| Adriatic (Gulf of Trieste) | 1.4-1.7 | 6.2-9.8 | 87-137 | 3.8-15 | | | | 167-219 | 0.28-1.3 | 3 |
| Aegean (Saronikos Gulf) | 0.06-0.08 | 4.5 | 12-87 | 83-110 | 39 | 0.11-7.8 | 0.0009-0.01 | 17-32 | 0.06-0.2 | 4, 5 |
| (Turkey) | 6.6-12 | 36-64 | 336-452 | | | 26-55 | | 308-356 | 0.89-1.1 | 6 |
| South-west Mediterranean (Algeria) | 0.3-6.5 | | 7.2-71 | | | | | | 0.25-0.63 | 7 |

Where necessary, values were converted using a wet/dry weight ratio of 6.

1. Fowler and Oregioni, 1976
2. Stoeppler <u>et al</u>., 1977
3. Marjori <u>et al</u>., 1979
4. Grimanis <u>et al</u>., 1979
5. Papadopoulou and Kanias. 1976
6. Uysal, 1979
7. Aissi, 1979

TABLE 22
PCB in marine samples from the Mediterranean Sea during 1974-1976
(Extract from MAP Technical Reports Series No. 28, table 23, Annex)

| Sample type | Date | Region | No. of samples | PCB (range) | PCB (X) |
|---|---|---|---|---|---|
| | | | | ng $l^{-1}$ | |
| Sea water | 10.74 | North-west Mediterranean | 11 | 1.5-38 | 13 |
| | 2.75 | Ligurian | 17 | 1.3-8.6 | 3.2 |
| | 2.75 | Aegean | 7 | 0.2-1.3 | 0.36 |
| | 5.75 | Ionian | 10 | 0.2-2.0 | 1.0 |
| | | Tyrrhenian and Algero-Provençal Basin | 34 | 0.2-5.9 | 2.0 |
| | 9.75 | Algero-Provençal Basin | 8 | 0.6-19 | 4.6 |
| | | | 7 | 0.6-4.8 | 2.5 |
| | | Tyrrhenian | 6 | 1.5-11.6 | 4.5 |
| | | | | ng $M3^{-1}$ | |
| Marine air | 8.75 to 1.76 | Monaco coast | 13 | 0.1-1.0 | 0.4 |
| | 1.76 to 2.76 | Monaco coast | 12 | 0.03-0.08 | 0.06 |
| | 9.75 | Algero-Provençal basin | 4 | 0.2-0.3 | 0.25 |
| | | Tyrrhenian | 2 | 0.1-0.3 | 0.2 |
| | | | | ng $kg^{-1}$ | |
| Sediments | 5.75 | Ionian | 3 | 0.8-5.1 | 2.8 |
| | | Algero-Provençal basin | 5 | 0.8-9.0 | 4.0 |
| | | Gibraltar sill and Siculo-Tunisian sill | 2 | 0.8 | 0.8 |
| | | Algerian marin | 1 | 9.0 | 9.0 |

The asymetric range about the mean indicates that "hot spots" were found in the surveys in the areas such as the coast from Toulon to Marseille, the Neapolitan harbour, the Gulf of Venice, the Pula area, the Rijeka Bay, the Saronicos Gulf, etc.

Biota: Means of PCBs available in marine organisms range from 1.5 to 815 ug PCBs $kg^{-1}$ FW. More numerous data are available only for the mussel and the red mullet. Both in mussel and in mullet MED POL area II seems to have the highest concentrations, but there must exist "hot spots" since maxima are very high despite relatively low means. In mullet the lowest level means and maxima are observed in areas IX and X. Means of pp DDT range from 0.1 to 343 ug $kg^{-1}$ FW with Thunnus thynnus having the highest levels. Again high maxima indicate "hot spots". pp DDD$^{-1}$ means range from 0.4 to 325 ug $kg^{-1}$ FW and pp DDE from 1.5 to 600 ug $kg^{-1}$ FW; tuna always presents the highest mean. Asymetric distribution of minima and maxima about the mean indicates a skewed distribution. Dieldrin means range from 0.4 to 6.2 ug $kg^{-1}$ FW, Aldrin means from 0.2 to 2 ug $kg^{-1}$ FW, Hexachloro/cychlohexane from 0.7 to 20 ug $kg^{-1}$ FW and Lindane from 0.4 to 19 ug $kg^{-1}$ FW. Where data are available for several areas and the same organism, area X seems to have the lowest mean and lowest maximum (Table 23).

TABLE 23
## Chlorinated hydrocarbons (ug Kg$^{-1}$ FW) in Mediterranean marine organisms
### (Extract from MAP Technical Reports Series No. 28, table 24, Annex)

| Region | Chlorinated Hydrocarbons | Species | No. of Samples | Mean concen. | Standard deviation | Range | |
|---|---|---|---|---|---|---|---|
| II | PCB | Mytilus galloprovincialis | 17 | 307 | 266 | 22 | - 1200 |
| IV | " | " " | 13 | 95 | 114 | 5 | - 420 |
| V | " | " " | 159 | 84 | 221 | 5 | - 2622 |
| VIII | " | " " | 12 | 62 | 12 | 40 | - 80 |
| II | PCB | Mullus barbatus | 33 | 813 | 1496 | 30 | - 8000 |
| IV | " | " | 33 | 417 | 770 | 50 | - 3950 |
| V | " | " | 86 | 234 | 473 | 1 | - 3117 |
| VIII | " | " | 51 | 113 | 204 | 0 | - 1110 |
| IX | " | " | 6 | 9.3 | 19 | 0,4 | - 52 |
| X | " | " | 42 | 69 | 75 | 0 | 284 |
| VIII | PCB | Parapenaeus longirostris | 30 | 12.3 | 12,2 | 0 | - 51 |
| IX | " | " " | 3 | 1.5 | - | 0 | - 2,5 |
| X | " | " " | 11 | 31 | 57 | 0 | - 157 |
| II | PCB | Carcinus mediterraneus | 10 | 12.3 | 12,2 | 0 | - 51 |
| V | " | " " | 3 | 1.5 | - | 0 | - 2,5 |
| X | " | | 11 | 31 | 57 | 0 | - 157 |
| IV | PCB | Mullus surmuletus | 6 | 87 | 17 | 60 | - 110 |
| V | " | " | 9 | 101 | 130 | 5 | - 441 |
| IV | PCB | Nephrops norvegicus | 28 | 25 | 17 | 8 | - 90 |
| II | pp DDT | Mullus barbatus | 27 | 28 | 35 | 8 | - 170 |
| IV | " | " | 33 | 23 | 17 | 6 | - 89 |
| V | " | " | 102 | 17 | 26 | 0,2 | - 205 |
| VIII | " | " | 51 | 23 | 25 | 4 | - 110 |
| IX | " | " | 17 | 38 | 29 | 0,5 | - 92 |
| X | " | " | 44 | 8 | 9 | 0 | - 37 |
| II | pp DDT | Mytilus galloprovincialis | 113 | 22 | 23 | 3 | - 150 |
| IV | " | " " | 12 | 7 | 5 | 1.2 | - 17 |
| VIII | " | " " | 180 | 15 | 77 | 0 | - 1014 |
| II | pp DDT | Thunnus thynnus thynnus | 21 | 343 | 362 | 25 | - 1401 |
| IV | pp DDT | Mullus surmuletus | 6 | 6 | 3 | 4 | - 13 |
| V | " | " | 11 | 9 | 11 | 0,5 | - 40 |
| V | pp DDT | Carcinus mediterraneus | 31 | 1.7 | 1.4 | 0.2 | - 5 |
| IX | " | " | 6 | 1.6 | 0.7 | 0.4 | - 2.6 |
| VIII | pp DDT | Parapenaeus longirostris | 29 | 0.9 | 1.4 | 0 | - 6 |
| II | " | " " | 4 | 4.2 | 3.5 | 0.3 | - 9 |
| X | " | | 10 | 0.1 | 0.2 | 0 | - 0.8 |
| II | Dieldrin | Mullus barbatus | 11 | 6.2 | 5.3 | 0.5 | - 19 |
| IV | " | " | 9 | 6 | 3.6 | 0.5 | - 12 |
| V | " | " | 67 | 1.7 | 4.1 | 0.1 | - 17 |
| X | " | " | 35 | 0.4 | 1.1 | 0 | - 35 |
| II | Dieldrin | Mytilus galloprovincialis | 2 | 3.5 | - | 1 | - 6 |
| IV | " | " | 6 | 2.5 | 2.6 | 0.5 | - 6 |
| V | " | " | 145 | 0.8 | 4.4 | 0.1 | - 56 |
| V | Dieldrin | Mullus surmuletus | 8 | 0.4 | 0.2 | 0 | - 0.7 |
| IV | Dieldrin | Nephros norvegicus | 7 | 0.9 | 0.5 | 0.5 | - 1.8 |
| V | Dieldrin | Carcinus mediterraneus | 31 | 0.5 | 0.6 | 0 | - 2.4 |
| X | " | | 4 | 3.1 | 4.5 | 0.4 | - 10 |

TABLE 23 (Cont'd./..2)
Chlorinated hydrocarbons (ug Kg$^{-1}$ FW) in Mediterranean marine organisms
(Extract from MAP Technical Reports Series No. 28, table 24, Annex)

| Region | Chlorinated Hydrocarbons | Species | No. of Samples | Mean concen. | Standard deviation | Range | | |
|---|---|---|---|---|---|---|---|---|
| II | Aldrin | Mullus barbatus | 9 | 0.5 | - | 0.5 | - | 0.5 |
| IV | " | " | 9 | 1.5 | 1.9 | 0.5 | - | 5 |
| IX | " | " | 5 | 0.5 | 0.4 | 0 | - | 1 |
| X | " | " | 44 | 1.5 | 4.7 | 0 | - | 28 |
| IV | Aldrin | Mytilus galloprovincialis | 6 | 2 | 2.1 | 0.5 | - | 5 |
| IV | Aldrin | Nephrosp norvegicus | 7 | 0.6 | 0.2 | 0.5 | - | 1 |
| X | Aldrin | Carcinus mediterraneus | 5 | 1.6 | 2.8 | 0 | - | 6.5 |
| IX | Aldrin | Parapenaeus longirostris | 4 | 1.4 | 1 | 0 | - | 2.8 |
| X | " | " | 11 | 0.2 | 0.6 | 0 | - | 2.2 |
| II | Hexachloro-cychlohexane | Mullus barbatus | 63 | 2.6 | 2.8 | 0.2 | - | 12 |
| VIII | " | " | 4 | 5 | 8 | 0.8 | - | 50 |
| IX | " | " | 5 | 3.9 | 3.9 | 1 | - | 11 |
| V | " | Mytilus galloprovincialis | 43 | 1.1 | 1 | 0 | - | 5 |
| VIII | " | " | 55 | 1.9 | 1.5 | 0.4 | - | 5 |
| V | " | Mullus surmuletus | 4 | 1.2 | 1.7 | 0 | - | 4 |
| V | " | Carcinus mediterraneus | 27 | 0.9 | - | 0 | - | 8 |
| IX | " | " | 6 | 20 | - | 12 | - | 34 |
| VIII | " | Parapenaeus longirostris | 7 | 0.7 | 0.3 | 0.2 | - | 1.1 |
| II | Lindane | Mullus barbatus | 17 | 19 | 14 | 2 | - | 36 |
| IV | " | " | 9 | 1.5 | 1.4 | 0.5 | - | 5 |
| V | " | " | 62 | 0.7 | 0.9 | 0 | - | 3.8 |
| II | " | Mytilus galloprovincialis | 7 | 4.8 | 6 | 0.5 | - | 20 |
| IV | " | " | 6 | 1.7 | 0.9 | 0.5 | - | 3 |
| V | " | " | 36 | 0.4 | 0.4 | 0 | - | 2 |
| II | Lindane | Carcinus mediterraneus | 4 | 19 | 14 | 2 | - | 36 |
| V | " | " | 27 | 0.2 | - | . | - | . |
| IV | " | Nephrops norvegicus | 7 | 0.5 | - | . | - | . |
| II | pp DDD $^{-1}$ | Mullus barbatus | 12 | 38 | 52 | 0 | - | 180 |
| V | " | " | 5 | 28 | 40 | 2.2 | - | 107 |
| VIII | " | " | 78 | 14 | 25 | 0 | - | 140 |
| IX | " | " | 17 | 18 | 14 | 0 | - | 44 |
| X | " | " | 44 | 1.6 | 3.8 | 0 | - | 21 |
| II | pp DDD $^{-1}$ | Mytilus galloprovincialis | 108 | 15 | 13 | 5 | - | 125 |
| V | " | " | 11 | 49 | 124 | 0 | - | 440 |
| VIII | " | " | 90 | 7 | 7 | 0 | - | 45 |
| II | pp DDD $^{-1}$ | Thunnus thynnus thynnus | 21 | 107 | 98 | 5 | - | 117 |
| VIII | " | " | 4 | 323 | 422 | 26 | - | 1052 |
| V | pp DDD $^{-1}$ | Mullus surmuletus | 3 | 7 | 6 | 2 | - | 15 |
| II | pp DDD $^{-1}$ | Carcinus mediterraneus | 10 | 10 | 9 | 1.2 | - | 26 |
| IX | " | " | 6 | 4.2 | 3.7 | 0 | - | 10 |
| VIII | pp DDD $^{-1}$ | Parapenaeus longirostris | 29 | 0.8 | 1.4 | 0 | - | 7 |
| IX | " | " | 4 | 2.2 | 1.3 | 0.5 | - | 4.2 |
| X | " | " | 11 | 0.4 | 0.8 | 0 | - | 2.7 |
| II | pp DDE $^{-1}$ | Mullus barbatus | 34 | 29 | 14 | 11 | - | 70 |
| IV | " | " | 33 | 33 | 18 | 7 | - | 93 |
| V | " | " | 43 | 8 | 12 | 0.1 | - | 75 |
| VIII | " | " | 88 | 33 | 39 | 1 | - | 255 |
| IX | " | " | 16 | 53 | 42 | 0,9 | - | 117 |
| X | " | " | 44 | 15 | 12 | 2 | - | 67 |

TABLE 23 (Cont'd./..3)
Chlorinated hydrocarbons (ug Kg$^{-1}$ FW) in Mediterranean marine organisms
(Extract from MAP Technical Reports Series No. 28, table 24, Annex)

| Region | Chlorinated Hydrocarbons | Species | No. of Samples | Mean concen. | Standard deviation | Range | | |
|---|---|---|---|---|---|---|---|---|
| II | pp DDE$^{-1}$ | Mytilus galloprovincialis | 114 | 13 | 9 | 2.2 | - | 42 |
| IV | " | " | 13 | 6 | 4 | 2 | - | 17 |
| V | " | " | 145 | 5 | 13 | 0.1 | - | 110 |
| VIII | " | " | 99 | 10 | 12 | 1 | - | 75 |
| II | pp DDE$^{-1}$ | Thunnusthynnus thynnus | 21 | 352 | 415 | 23 | - | 1582 |
| VIII | " | " | 4 | 601 | 659 | 161 | - | 1737 |
| IV | pp DDE$^{-1}$ | Mullus surmuletus | 6 | 11 | 3 | 6 | - | 15 |
| V | " | " | 10 | 12 | 12 | 0.1 | - | 33 |
| II | pp DDE$^{-1}$ | Carcinus mediterraneus | 10 | 36 | 24 | 14 | - | 72 |
| V | " | " | 4 | 2.5 | 30 | 0.1 | - | 6.2 |
| VIII | " | " | 3 | 23 | 3 | 20 | - | 26 |
| IX | " | " | 7 | 22 | 15 | 0.3 | - | 45 |
| X | " | " | 4 | 3.1 | 3.5 | 0.7 | - | 8 |
| IV | pp DDE$^{-1}$ | Nephrops norvegicus | 28 | 3.8 | 1.8 | 1.1 | - | 8 |
| VIII | pp DDE$^{-1}$ | Parapenaeus longirostris | 31 | 1.6 | 5 | 0 | - | 25 |
| IX | " | " | 4 | 3.1 | 1.6 | 1 | - | 5.4 |
| X | " | " | 11 | 1.5 | 2.6 | 0 | - | 9 |

## Petroleum Hydrocarbons

Sea water:  In general, dissolved/dispersed petroleum
hydrocarbons (DDPH) in water samples collected off-shore have
concentrations below 10 ug DDPH l$^{-1}$, but very high concentrations have
also been found occasionally in off-shore waters (Table 24).
GC-determination on open sea samples (Phycemed cruises) showed that the
petrogenic hydrocarbons' aliphatic fraction ranged between 1.1. and 4.5
ug l$^{-1}$ and the aromatic one from 0.1 to 0.8 ug l$^{-1}$.  For instance, between
Castellon and Cartagena in the Western Mediterranean along 9 transects,
station means varied between 0.06 and 8.25 ug l$^{-1}$.  Near the shore,
particularly near industrialized areas or river mouths, concentrations
can be well above 10 ug DDPH l$^{-1}$. In the inner part of the harbour of
Taranto (Mar Piccolo) concentrations varied between 0.1 and 36 ug l$^{-1}$.
The data from the Adriatic refer only to the Rijeka Bay, Sibenik and
Split.  In unpolluted areas, values are 0.5 ug l$^{-1}$ or below.  In polluted
areas, up to 50 ug l$^{-1}$ have been determined (IR determinations gave much
higher concentrations: up to 1100 ug l$^{-1}$). In the Central Mediterranean,
recent offshore values ranged from 9.3 to 28 ug l$^{-1}$, but older data had
maxima up to 425 ug l$^{-1}$.  Recent nearshore values had concentration
ranges very similar to the recent offshore concentrations (Libyan coast).
In the Eastern Mediterranean along the Greek coasts, concentrations
varied from 0.1 to 2.6 ug l$^{-1}$ with the highest concentrations in harbour
areas. In the Aegean Sea concentrations ranged from 0.15 to 1.4 ug l$^{-1}$,
but with some "hot spots" far from major land-based sources with levels
up to over 10 ug l$^{-1}$. In Greek harbours very high IR data have been
obtained. Also along the Turkish coast concentrations varied over a wide
range (0.02 to 40 ug l$^{-1}$), with high off-shore concentrations most
probably caused by direct discharges from ships. In Israel, near
harbours, oil refineries, river mouths, etc. 10 to 20 ug l$^{-1}$ were
reported.  Near Cyprus, concentrations ranged from 25 to 40 ug l$^{-1}$.
Similar high concentrations were also determined in coastal waters off
Egypt.  Obviously, discharges from ships can greatly influence the
concentrations in offshore waters [30].  In general, it appears that
recent concentrations are lower, but the data so far available are
insufficient for a statistical treatment.

TABLE 24
Dissolved/dispersed petroleum hydrocarbons(ug $1^{-1}$)
(Extract from MAP Technical Reports Series No. 28, table 25, Annex)

| Area | Year | Concentrations | Technique | Reference |
|---|---|---|---|---|
| **Western Mediterranean** (offshore) | | | | |
| Northern part | 1973 | 10-2200 (surface) (av.450) 3-37 (10m) (av. 15) | Fluorescence | Monaghan et.al., 1974 |
| | 1975-77 | 2-6 (surface) (av. 3.3) | " | Faraco and Ros, 1978 |
| | 1981 | 1.5-21.1 (surface) 3.5-4.6 (surface) 0.5-0.8 (chr. eq.) | GC-n-alkanes -UCM fluorescence | Ho et al., 1982 " " |
| | 1983 | 1.9 (surface) 1.3 (surface) | GC-n-alkanes -UCM | Sicre et al., 1984 |
| Central part | 1981 | 0.33 (chr. eq.) | Fluorescence | Ho et al., 1982 |
| | 1983 | 0.68 (surface) 1.37 (surface) | GC-n-alkanes -UCM | Sicre et al., 1984 |
| Southern part | 1973 | 2-17 (surface) (av. 8.5) 2.7 (10 m) | Fluorescence | Monaghan et al., 1974 |
| | 1974-75 | av. 6.9 (surface) | " | Zsolnay, 1979 |
| | 1975-77 | 1-123.5 (surface) (av. 17.5) | " | Faraco and Ros, 1978 |
| | 1981 | 0.23 (surface) 0.81 (surface) 0.078-0.2 (chr. eq.) | GC-n-alkanes -UCM Fluorescence | Ho et al., 1982 " " |
| | 1983 | 0.31 (surface) 1.15 (surface) | GC-n-alkanes -UCM | Sicre et al., 1984 |
| Alboran Sea | 1975-77 | 4.3-14.6 (surface) (av. 7.9) | Fluorescence | Faraco and Ros, 1978 |
| | 1981 | 0.2 (chr. eq.) | " | Ho et al., 1982 |
| **Western Mediterranean** (nearshore) | | | | |
| Spanish coast | | | | |
| Castellon | 1983 | 1.36-2.40 | " | De Leon, 1984 |
| Sagunto | " | 0.06-3.40 | " | " |
| Valencia | " | 0.63-4.35 | " | " |
| Cullera | " | 0.06-3.10 | " | De Leon, pers.comm. |
| Benidorm | " | 0.60-0.26 | " | " |
| Alicante | " | 0.85-8.26 | " | |
| Guardamar | " | 1.15-3.15 | " | |
| Portman | " | 0.26-6.50 | " | |
| Cartagena | " | 0.26-3.22 | " | |
| French coast | | | | |
| Banyuls-sur-Mer Var Estuary | 1975-78 | 50-5000 (av. 580) | IR | UNEP, 1980 |
| Gulf of Fos | 1981 | 0.4-1.0 | GC-UCM | Burns and Villeneuve, 1982 |
| Gulf of Ajaccio | 1983-84 | 30-200 | | MEDPOL Phase II |
| | 1983-84 | 0-100 | | MEDPOL Phase II |

TABLE 24 (Cont'd./..2)
Dissolved/dispersed petroleum hydrocarbons(ug $1^{-1}$)
(Extract from MAP Technical Reports Series No. 28, table 25, Annex)

| Area | Year | Concentrations | Technique | Reference |
|---|---|---|---|---|
| Italian coast | | | | |
| Tyrrhenian Sea | 1973 | 8-614 (surface) (av. 180) 3-19 (10 m) (av. 7) | Fluorescence | Monaghan et al., 1974 |
| | 1974-75 | av. 4.8 (surface) | " | Zsolnay, 1979 |
| | 1975-77 | 1.9-20.5 (av. 7.4) | " | Faraco and Ros, 1978 |
| Taranto, Mar Piccolo | 1983 | 0.2-11.6 (av. 3.26) | GC | Strusi, pers. comm. |
| " | " | 0.5-23.0 (av. 7.42) | " | " |
| " | " | 0.1-36.0 (av. 7.98) | " | " |
| **Central Mediterranean** | | | | |
| South Ionian Sea | 1973 | 3-423 (surface) (av. 58) 2-120 (10 m) (av. 16) | Fluorescence | Monaghan et al., 1974 " " |
| | 1974-75 | av. 14.9 (surface) | " | Zsolnay, 1979 |
| Malta, coastal waters | 1977-78 | 0.02-0.29 | " | UNEP, 1980 |
| | 1984 | 0.03-1.70 (av. 0.51) | " | UNEP, 1985 (MEDPOL Phase II) |
| Libyan coast | 1974-75 | av. 24.9 (surface | Fluorescence | Zsolnay, 1979 |
| W Sedra, Tripoli harbour | | | | |
| Zawia | 1980 | 20-28 | " | Gerges and Durgham, 1982 |
| Janzur, W&E Brega, Zawia | | | | |
| W Khoms | " | 12.5-19 | " | " |
| Zlitan, Zwetina,Benghazi, | | | | |
| E Sirte, Tajura | | 4.6-5.3 | " | " |
| Sabratha, Derma, Sidi Blal | | 0.6-2.9 | " | |
| Libyan coast 171 samples from coastal areas | " | 0.0-27.6 (av. 3.6) | " | MFRC, Tripoli 1981 |
| Adriatic | | | | |
| Yugoslavia, Rijeka Bay | 1976-77 | 1-50 | Fluorescence | UNEP, 1980 |
| | | 100-1100 | IR | |
| | | below 0.1 | GC | Ahel & Picer, 1978 |
| " " | 1976-78 | 1-7 ("polluted") | Fluorescence | Ahel, 1984 |
| | | 0.2-0.5 ("unpolluted") | " | |
| Yugoslavia, Sibernik area | 1984 | 0.2-16.4 (av. 1.4) | " | UNEP, 1985 (MEDPOL Phase II) |
| Yugoslavia, Split | 1984 | av. 24.9 (surface) | " | " " |
| **Eastern Mediterranean** | | | | |
| Aegean Sea | 1974-75 | av. 20.5 | Fluorescence | Zsolnay, 1979 |
| Greece | | | | |
| Coastalwaters | | below 3 | " | Mimicos, 1980 |
| Saronikos Gulf | 1980-81 | 1.6-5.6 | " | Gabrielides et al., 1984 |
| Aegean Sea | " | 2.9-13.7 | " | " |
| Thessaloniki harbour | 1976-79 | 1500 | IR | UNEP, 1980 |
| Cavala harbour | " | 2600 | " | " |
| Strymonikos Bay | " | 1100 | " | " |
| Patraikos Gulf | 1977-83 | 0.12-28.2 | Fluorescence | Mimicos et al., 1984 |
| Acheloos River estuary | " | 1.3-4.5 | " | " |

TABLE 24 (Cont'd./..3)
Dissolved/dispersed petroleum hydrocarbons(ug  l $^{-1}$)
(Extract from MAP Technical Reports Series No. 28, table 25, Annex)

| Area | Year | Concentrations | Technique | Reference |
|---|---|---|---|---|
| **Turkey** | | | | |
| Mersin-Akkuyu | 1977-78 | 8.2-39.4 | Fluorescence | UNEP, 1980 |
| Southern coast | 1980-82 | 0.5-3.5 | " | Sunay et al., 1982 |
| | | (av. 1.5) | | |
| **Offshore between Turkey** | | | | |
| and Cyprus | 1980-82 | 2.6-6.0 | " | Sunay et al., 1982 |
| Iskenderun Bay | " | 0.7-7.0 | " | |
| Sea of Marmara | 1983 | 0.88(max. 8.07) | " | Sakarya et al., 1984 |
| Izmit Bay | " | 0.75-5.0 | " | " |
| Aegean Sea | " | 0.14-1.39 | " | " |
| Mediterr.coastal waters | " | 0.02-1.1 | " | " |
| Iskenderun Bay | " | 0.11-1.0 | " | " |
| Candarli Bay | 1983-84 | 1.20-80.0 | " | Topcu & Muezzinoglu, 1984 |
| Aliaga | " | 0.53-7.30 | " | " |
| Saros Bay | 1983 | 0.77 | " | " |
| Izmir Bay | " | 9.40 | " | " |
| Southern Aegean Coast | " | 0.86 | " | " |
| **Eastern Mediterranean** | | | | |
| **Offshore South of Cyprus,** | | | | |
| Southeast of Crete | 1975-76 | 10-40 | " | UNEP, 1980 |
| **Israel** | | | | |
| Ashkelon | 1975-76 | 9.4-19.4 | Fluorescence | UNEP, 1980 |
| Haifa Bay | " | 15.0-15.6 | " | " |
| Plamachin | " | 10.7-12.5 | " | " |
| Bardawil Lagoon | " | 20.6 | " | " |
| Tel Shikmona | " | 1.1-45.3 | " | " |
| **Egypt** | | | | |
| Alexandria | 1978-79 | 0.7-35.2 (av. 3.7) | " | Aboul-Dahab & Halim, 1980 |
| Alexandria | 1979-80 | 6.6-41.4 | " | Wahby and El Deeb, 1980 |
| | | (nearshore) | " | " |
| Mouth of Suez Canal | 1980-81 | 0.7-3.9 (offshore) | " | Samra et al., 1982 |
| | | 0.5-14 | | |
| Cyprus, Limassol Bay | 1983 | 2.6-8.1 | Fluorescence | MEDPOL Phase II |
| | 1984 | 1.15-1.48 | " | " |
| Larnaca Bay | 1983 | 4.2-13.6 | " | " |
| | 1984 | 1.74-2.53 | " | " |

The limited data on floating tar show arithmetic mean levels ranging from 0.5 to 16 mg m$^{-2}$ in offshore areas and much higher concentrations in nearshore waters (10 to 100 mg m$^{-2}$). The Eastern basin appears most heavily contaminated. Also tar on beaches was relatively high but only data for the Eastern Mediterranean exist [33].

Sediments: Petroleum hydrocarbon data in sediments range from 1 to 62 ug aliphatics g$^{-1}$ sediment and 2 to 66 ug aromatics g$^{-1}$ sediment along the Spanish coast outside harbours, oil terminals and river mouths. Two samples from the central part of the western basin suggest background levels in the order of 1.2 ug g$^{-1}$ aliphatics and 0.6 ug g$^{-1}$ aromatics. Offshore sediments from Cyprus (90 m depth) ranged from 0.114 to 1.35 ug g$^{-1}$ and somewhat lower values are reported from Iskenderun Bay, Turkey.

Too few data are available to obtain a distribution pattern of petroleum hydrocarbons in the Mediterranean sediments [33].

Organisms:  Only very few studies on petroleum hydrocarbons in marine organisms are available. Mussels collected near Palamos, Barcelona and in the Ebro Delta (8 to 3200 ug g$^{-1}$ DW in the saturate fraction) are much higher than in fishes. Mullus barbatus (5.8 to 22 ug g$^{-1}$) has somewhat higher concentrations than pelagic fishes such as Merluccius sp. Trachurus sp. and Engraulis encrasicholus [33].

Microbiological Quality of Recreational Waters and Shell-Fish growing Waters

Guidelines for monitoring the quality of coastal recreational and shellfish areas were prepared by WHO within the framework of the MED POL Phase II programme, and issued by UNEP's Regional Seas Programme as part of its Series of Reference Methods for Marine Pollution Studies.  For coastal recreational waters, as a minimum programme, the guidelines recommend faecal coliforms, faecal streptococci and at least one pathogen causing infection through contact, together with complementary observations, principally salinity and temperature, as the constituent parameters of minimum monitoring programmes.

For shellfish areas, the guidelines recommend faecal coliforms in the shellfish themselves as the main parameter, with weekly sampling during the peak consumption period in addition to the 3-month frequency laid down as a minimum.  The guidelines also recommend that the water in the shellfish-growing areas be monitored (at the same frequency) for faecal coliforms, faecal streptococci and any pathogen considered important in the light of local requirements from time to time.

Sanitary surveillance of beaches is considered as an integral part of coastal recreational water quality programmes, as a significant amount of time is spent on the beach itself.  Apart from the utilization of a classification system for the visual appearance of beaches, according to the McKay procedure (a modified version of the Garber classification), the guidelines also recommend microbiological examination of beach sand for the presence of faecal coliforms, faecal streptococci and fungi.

Since 1982, considerable effort has been directed towards the establishment or enhancement of national marine pollution monitoring programmes within the framework of the MED POL - PHASE II programme, which would include monitoring of microbial pollution.

At present, all Mediterranean countries possess a marine pollution monitoring programme covering all or part of the main coastal recreational and (to a lesser extent) shellfish areas.  An evaluation of the Monitoring data for recreational waters submitted by seven countries (Algeria, Cyprus, Israel, Lebanon, Malta, Morocco and Yugoslavia) for the period 1983-1987 was performed in 1989 [34].  According to this evaluation (table 25), the conformity of sampling stations with the interim criteria adopted by Mediterranean Governments on a common basis in 1985 rose from 78% (1983) to 96% (1987) in the case of stations with at least six samplings per year.  In this regard, although the number of stations monitored differed slightly from year to year, the figures can be considered as providing a reasonable evaluation of the positive trend in the countries in question, taken as a whole.

Information is also available from reports of EEC Member States (France, Greece, Italy and Spain) in terms of the 1976 bathing water quality Directive.  In France the number of monitoring stations with high to acceptable quality water (A, AB or B) rose from 76.4% (1983) to 83% (1987), with a corresponding reduction in lower quality stations (23.6% in 1983 to 16.7% in 1987).  In Italy, the number of stations conforming with Italian criteria (based on, but stricter than, the Directive) rose from 68% in 1984 to 87.3% in 1988.  In Spain, the number of stations with high quality water (A2) decreased from 65.2% in 1986 to 51.0% in 1987. Results from Greece are only available for 1987, and are restricted to Attica, where 77.7% of stations with at least 5 sampling per year were found to be in conformity with requirements.

The information available with respect to the eleven countries above mentioned is sufficient to indicate a general trend of improvement during the last five years over a stretch of the Mediterranean coastline

## TABLE 25
Preliminary evaluation of the microbiological quality of recreational
waters in the Mediterranean according to the interim Mediterranean
environmental quality criteria.  Sampling stations with at
least 6 samples per year.  MED POL Phase II: 1983-1987.
(Extract from document UNEP(OCA)/MED WG.5/Inf.4)

| Year | Samples analyzed | Stations surveyed | Average sampling frequency | Sampling stations conforming to | | |
|------|------|------|------|------|------|------|
| | | | | FC50 | FC90 | FC50 & FC90 |
| 1983 | 524 | 50 | 10 | 43 (86%) | 39 (78%) | 39 (78%) |
| 1984 | 1755 | 133 | 13 | 120 (90%) | 111 (83%) | 108 (81%) |
| 1985 | 2178 | 128 | 17 | 115 (90%) | 104 (81%) | 102 (80%) |
| 1986 | 3048 | 238 | 13 | 216 (92%) | 200 (84%) | 200 (84%) |
| 1987 | 1908 | 150 | 13 | 145 (97%) | 145 (97%) | 144 (96%) |
| Total | 9413 | 699 | 15 | 639 (91%) | 599 (86%) | 593 (85%) |

covering practically the whole of the northern seaboard, the western part
of the southern, and part of the central and eastern areas.
    The situation can perhaps be described as having regressed slightly
since the original Coastal Water Quality Control pilot project
operational from 1976 to 1980 in that, during the course of that project,
practically all participating laboratories measured at least three major
bacterial parameters - total coliforms, faecal coliforms and faecal
streptococci.  Since 1982, the majority of countries with national
monitoring agreements have restricted their microbiological parameters to
faecal coliforms, a practice which has assumed a cloak of legality since
the adoption of the joint interim criteria for bathing waters by
Mediterranean Governments in 1985.
    A considerable amount of data has been obtained since the launching
of the MED POL - PHASE II programme in 1982, particularly with regard to
the adaptation and survival of micro-organisms in the sea, less so in the
case of halogen-indicator relationship.  Although a number of
epidemiological studies have been performed and more are ongoing, most of
these are necessarily small-scale, and difficulties have been encountered
both from the financial and technical viewpoints, the former because of
lack of funds and the latter because of lack of common acceptable
designs.
    The situation regarding shellfish requires particular attention.
The main shellfish-producing countries in the region have their own
monitoring programmes, carried out under the terms of relatively strict
legislation.  However, very few results have been reported from the
countries carrying out monitoring programme through signed MED POL
agreements, with the result that the state of shellfish waters in the
region is very much less known than that of recreational waters.  In view
of the fact that the majority of intestinal pathogens connected to
seawater pollution are transmitted to man through shellfish consumption,
solution of this particular problem is of major importance.

**Eutrophication and Associated Phenomena**
The term "eutrophication" is borrowed from the limnological terminology.
In the marine environment, the term in its broad sense is used to
designate the primary and secondary impacts of excessive enrichment with
nutrient salts.  The concept should only apply to those marine areas
showing the following symptoms:

-    the rate of plant production (macro-algae, phytoplankton) exceeds
     that of consumption by herbivores;
-    dissolved oxygen is abnormally low or depleted, often with
     formation of hydrogen sulphide;
-    the ecosystem structure is disrupted, showing abnormal diversity
     and dominance indices and a change in the species composition.

Land-based sources discharging of nutrients into the Mediterranean basin are of three types, river input, agricultural run-off, and domestic-industrial waste water input. The respective impacts of river input and domestic industrial waste water inputs on the ecosystem are basically different. River discharge, when not in a confined coastal zone, usually spreads over vast sea areas, enhancing productivity, while becoming gradually diluted. Domestic-industrial waste water contributes a negligible 2% to the total fresh-water input to the Mediterranean basin, but its concentration in nutrients is one order of magnitude higher. Domestic-industrial effluents appear to contribute about 20% and 17% respectively to the total nitrogen and phosphorus inputs. Domestic waste waters in the Mediterranean usually remain restricted to the vicinity of the outlets of effluents, their heavy load of suspended materials rapidly settling to the bottom. Their impact on both pelagic and benthic ecosystems, therefore, is more localized but more severe.

Eutrophication, as reported for embayments and estuaries around the Mediterranean, is mostly associated with the release of untreated domestic-industrial waste water [35]. The symptoms range from mild to severe: alterations of the ecosystem structure involving the biomass, the specific diversity, the dominance of indicator species, outbursts of heavy, often almost monospecific, blooms or "red tides", the development of anoxic and azoic conditions. Such symptoms are either sporadic or seasonally recurrent. Some representative examples are outlined below.

The "lac de Tunis", a salt water lagoon receiving municipal water from the city of Tunis, exhibits extreme symptoms of eutrophication in summer. Massive growth of the chlorophyte Ulva covers a third of the lagoon surface, and the water column (1m in depth) becomes anaerobic, with blooms of red sulfur-reducing bacteria. A vigorous release of hydrogen-sulphide follows and fish-kills may amount to 10% of the total lagoon yield. Nearly a third of the lagoon is occupied by the reef-building, detritus-feeder polychaete Ficopomatus (Mercierella) enigmatica. The reefs harbour epizoan invertebrates, epiphytic algae and a rich microbial flora. This association releases large amounts of ammonia to the water. The heavy discharge of nutrients and organics from the city accompanied by a rapid turnover of nitrogen and a very slow flushing rate, create one of the most severely eutrophied Mediterranean environments [36].

The north Adriatic environment is predominantly governed by the discharge of the river Po and a number of smaller rivers, approximating 15% of the total fresh-water input to the Mediterranean basin. Heavy phytoplankton blooms are historically known to occur in the North Adriatic ("mare sporco") during spring or autumn. Severe eutrophic conditions, however, appear to be restricted to semi-closed bays receiving domestic waste water and to the area downstream from the Po outlet, along the Emilia-Romagna coast. The Central and South Adriatic remain poorer in nutrients than the Ionian Sea [37], [38] and [39].

On the other hand, the increased nutritional input into the North Adriatic has had beneficial effects on some commercially exploited species. During the last 20 years, the mediolittoral and the infra-littoral along the entire coast of the Istrian peninsula became covered with dense populations of the filter-feeding shellfish Mytilus sp. and Crassostrea sp. The stocks of small pelagic fish (anchovy and pilchard) are steadily increasing per unit fishing effort [38].

Kastela Bay, which receives waste waters from Split in the North-east Adriatic, has developed distinct symptoms of eutrophication. Primary productivity nearly doubled from 1962 to 1977 (from 115 to 206 g $C m^{-2}$) and the phytoplankton standing crop increased by a factor of ten in the meantime. Its composition shows the dominance of small diatom species which were inexistant before 1972: Nitzschia seriata, Skeletonema costatum and Leptocylindrus danicus. In summer 1980, a red tide of Gonyaulax polyedra reached $18 \times 10^6$ cells $l^{-1}$. This outbreak lasted for a few days and was followed by a near anoxicity in the bottom water layer and mass mortality of fish [40].

Pronounced summer eutrophication is recurrent in Elefsis Bay in the North of Saronikos Gulf near Athens. This is the result of the impact of the Keratsini outfall, draining domestic waste water from the city, and from thermohaline stratification. The Bay remains strongly stratified

from June to October.  In winter, dissolved oxygen is relatively high in
the bottom water, dropping to depletion in summer.  The average ratio of
total nutrients to background open water values is slightly over 10.  An
intense growth of diatoms and dinoflagellates develops in the Bay,
abruptly declining towards the open sea.  The zooplankton biomass drops
to zero from August to October, recovering the normal values in January-
April.  _Acartia clausi_ makes up 90-99% of the population.  A parallel
seasonal trend is shown by the benthic macrofauna.  In summer, the
central azoic zone of the bay widens in area and the benthic
invertebrates become restricted to the shallow coastal belt [41], [42].

     A distinction must be made between abnormally dense multispecies
phytoplankton blooms and "red tides", caused by the proliferation of a
dinoflagellate species.  Several features confirm the significance of
"red tides" and abnormal blooms as symptoms of eutrophication.  In all
instances, such abnormally dense outbreaks can only be sustained by
abnormally massive inputs of nutrient salts and organics.  The outbreaks
are triggered by the onset of stable stratification, quiet weather and
rising temperature.  Most of the reported cases from the Mediterranean
are recurrent phenomena, reappearing in the same season, at the same
localities in successive years, lasting from a few days to several weeks.

     There are abundant records of "red tides" along the coasts of the
Mediterranean basin (Table 26): from Castellon and the coast of Spain,
from Banyuls/mer and Juan-les-Pins in France, from the coast of Emilia-
Romagna and Kastela Bay in the North Adriatic, from Izmir Gulf in Turkey,
from the East harbour in Alexandria, Egypt, and from Malta.  Abnormally
heavy blooms due to mixed populations occur at the Rhône and Po outlets;
they also used to occur along the Nile delta.  Chloromonadine species are
also responsible for some of the outbreaks.  Similar outbreaks occur in
salt and brackish lagoons and estuaries, such as the lagoon of Venice and
several other Italian lagoons, the "lac de Tunis", the lakes on the
Island of Mljet in the Adriatic and the two Nile estuaries.

     Secondary effects follow the outbreak and subsidence of red tides
and abnormal blooms.  The sinking of massive amounts of senescent cells
that follows and their subsequent decay in the deeper water layer result
in a rapid depletion in dissolved oxygen.  Fish mortality, following the
development of red tides has been reported from some Mediterranean
localities:  Izmir Bay, Kastela Bay, Emilia-Romagna, Tunis lagoon.  Since
no transmissible toxins through the food chain have been reported, it is
likely that such mortalities result from oxygen depletion in semi-closed
embayments, the release of $H_2S$ and possibly from the smothering of fish
gills by the sinking cells.

     Eutrophication is for the time being a local rather than a regional
problem in the Mediterranean Sea.  It is frequent in the North Adriatic,
Izmir Bay, Elefsis Bay, the lagoon of Tunis and in all such areas where
the rate of input of domestic-industrial waste water exceeds that of the
exchange with the open sea.  The main symptoms consist in the disruption
of the community structure, the excessive macro- and micro-algal
production and the development of anoxicity in the subsurface layer.  The
open sea oligotrophic Mediterranean waters, however, still remain in
sharp contrast with the conditions prevailing in its peripheral
embayments.

## Radioactive Substances

Excluding natural radionuclides which occur everywhere on the earth's
surface as well as in its interior, the anthropogenic input of
radioactive substances into the marine environment has been restricted
and controlled, except in the case of nuclear weapon testing and
accidents.  In fact, radioactive substances resulting from nuclear
installations are rare pollutants on which severe control and monitoring
of their releases has been exercised, although their data has not always
been published.

     The largest quantities of radioactive wastes are, in general,
produced during the nuclear fuel reprocessing, followed by those in
nuclear power production, but at much lower levels; those from research
reactors, hospitals, etc. are normally negligible.  Since there is no
operating fuel reprocessing plant on an industrial scale in the

TABLE 26
Red tides in the Mediterranean Sea
(Dino=Dinoflagellate, Chloro=Chloromonadina, Diato=Diatoms)
(Extract from MAP Technical Reports Series No. 28, table 75, Annex)

| Locality | Biomass $10^6$ cell. $1^{-1}$ | Sporadic (+) Recurrent (++) | Organisms |
|---|---|---|---|
| Barcelona harbour | 1-52 | ++ | Chloro |
| Castellon | 21 | ? | Dino |
| Coast of Catalonia | 0.38 | ++ | Dino |
| Banyuls-sur-mer | 2 | ++ | Dino |
| Rhône outlet | - | ++ | Dino, Diato |
| Juan-les-Pins | 80 | ++ | Dino |
| Bay of Antibes | - | + | Chloro |
| Villefranche-sur-mer | - | ++ | Chloro |
| Emilia-Romagna | 10-100 | ++ | Dino, Diato |
| Kastela Bay | 18 | ? | Dino |
| Izmir Gulf | - | ++ | Dino |
| Alexandria, East Harbour | 26 | ++ | Dino |
| Algiers harbour | - | ? | Chloro |
| Malta | 0.4 | ++ | Dino |
| Messina | - | ? | Chloro |

Mediterranean region, most of the intended radioactive releases into the marine environment have originated from nuclear power plants, which are in operation only in four Mediterranean countries on its northern coast (France, Italy, Spain and Yugoslavia). Most of these power plants (approximately 25 units) are located along large rivers, such as the Rhône, Po, etc., so that radioactive effluents released from these installations are transported to the Mediterranean by the various river systems. It has been estimated, however, that the river input of artificial radionuclides, including soil erosion, is approximately 4% of the total input during the pre-Chernobyl period, the major input being that through wet/dry atmospheric fallout originated from nuclear explosion tests [43]. Thus, the anthropogenic radioactive releases from nuclear installations constitute only a minor source of artificial radioactivity in the Mediterranean, except for some special local situations.

The atmospheric fallout (wet and dry) resulting from the Chernobyl accident significantly affected the Mediterranean inventories of some long-lived radionuclides like $^{137}Cs$, etc., but not those of others. Most of the radionuclides brought into the Mediterranean region by the Chernobyl fallout were short-lived fission products such as $^{95}Zr$, $^{95}Nb$, $^{103}Ru$, $^{132}Te$, $^{131}I$, $^{141}Ce$, etc. which decayed quickly or in the course of 1986. Even the levels of medium-lived radionuclides such as $^{110m}Ag$, $^{106}Ru$, $^{134}Cs$, $^{141}Ce$, etc. in marine environmental matrices became very low due to their dispersion within the environment and their physical decay during the same period [44]. In 1988 only a few radionuclides resulting from the Chernobyl accident remain measurable in the marine environment.

Dumping of radioactive substances is not foreseen for the Mediterranean Sea and the release of liquid low-level radioactive wastes is severely restricted hence only very small amounts will reach the marine environment from normal operations of nuclear reactors installed near the coast and those which release radionuclides into rivers discharging into the Mediterranean.

In order to obtain further information on the behaviour and fate of radionuclides introduced into the marine environment, the existing control and monitoring of radioactive releases should be continued and, if necessary, strengthened in the future depending on the development of nuclear industry in the Mediterranean region.

## CONCLUSIONS

There is no doubt of the success in the implementation of the various components of the Mediterranean Action Plan but there is general agreement in attributing the greatest achievements to the environmental assessment component (MED POL).

The activities carried out during MED POL - PHASE I substantially contributed to the improvement of the technical capabilities of participating institutes through training and provision of equipment. MED POL helped to lay the foundations of a Mediterranean scientific infrastructure by contributing to the equipping of national laboratories, encouraging scientists to study their marine environment and enabling them to collect objective information and data on the quality of the Mediterranean environment. MED POL - PHASE II, in spite of some initial difficulties, is now fully operational. By the end of August 1991 all coastal States - including Albania who became a Contracting Parties to the Barcelona Convention in 1990 - were contributing to MED POL through ongoing Monitoring Programmes. Reports on the results of those programmes are received regularly from most countries. As to the research component, by the end of 1991, there were about 100 ongoing research projects being carried out in 14 Mediterranean countries.

The implementation of the various MED POL activities underlined the importance of having common standardized methods and reporting formats followed by all participating research centres as well as the great usefulness of intercalibration exercises and data quality assurance programmes. The efforts undertaken by MED POL in this respect were largely repaid by the fact that Mediterranean data are now numerous, easy to be interpreted and reliable [4].

Work to determine the state of pollution of the Mediterranean Sea has been extensive and is continuing through MED POL as part of the workplan for the implementation of the Land-based Sources Protocol. According to the Protocol, the pollution by the substances listed in annexes I and II of the Protocol has to be assessed and, accordingly, control measures have to be proposed to Contracting Parties for adoption (table 27). As a result, by the end of 1991 pollution assessment documents had been prepared as follows: Microbial Quality of Bathing Waters [45], Mercury [46], Microbial Quality of Shell-Fish waters [47], Used Lubricating Oils [48], Cadmium and Cadmium Compounds [49], Organotin Compounds [50], Organohalogen Compounds [51], Organophosphorus Compounds [52], Persistent Synthetic Materials [53], Radioactive Substances [54], Pathogenic micro-organisms [55], and the relative control measures have been adopted by Contracting Parties [56],[57]. It is worth mentioning that the data used for the preparation of those documents have mostly come from the MED POL research and monitoring activities.

However, the MED POL Programme did not show free from weaknesses. The dual inter-governmental and interdisciplinary approach, and also the degree of complexity and geographical scale of the studies has led to inevitable difficulties which are normally associated with these kind of activities. Also, on the one hand the resources required by a full-scale implementation of MED POL are definitely higher than the MED POL funds approved by the Governments and on the other hand correspondent resources have not been always foreseen and met by national authorities, particularly as regards vessel support and number and continuity of scientific staff.

However, on the whole the MED POL Programme has successfully acted as catalyst on the subject of marine pollution studies and, together with non-governmental organizations, has increased the public awareness on the subject of marine pollution; MED POL has also identified the major environmental problems of the Mediterranean, has improved the capabilities of the Mediterranean scientific community and now, through the gradual implementation of the Land-based Sources Protocol and the provision of pollution control measures, offers to Mediterranean Countries the tools for a proper environmentally sound development.

TABLE 27
Workplan for the preparation of pollution assessment documents including
control measures according to the substances of annex I and annex II to
the Land-based Sourses Protocol

| | Substance (from annex I or annex II of Land-based Sources Protocol) | Year |
|---|---|---|
| 1. | Used Lubricating Oils | 1986 |
| 2. | Shell-Fish and Shell-Fish Growing Waters | 1986 |
| 3. | Cadmium and Cadmium Compounds | 1987 |
| 4. | Mercury and Mercury Compounds | 1987 |
| 5. | Organohalogen Compounds | 1987 |
| 6. | Persistent Synthetic Materials which may float, sink or remain in suspension | 1988 |
| 7. | Organophosphorous Compounds | 1988 |
| 8. | Organotin Compounds | 1988 |
| 9. | Radioactive Substances | 1989 |
| 10. | Carcinogenic, Teratogenic or Mutagenic Substances | 1989 |
| 11. | Pathogenic Micro-organisms | 1989 |
| 12. | Crude Oils and Hydrocarbons of any Origin | 1990 |
| 13. | Zinc, Copper and Lead | 1990 |
| 14. | Nickel, Chromium, Selenium and Arsenic | 1990 |
| 15. | Inorganic Compounds of Phosphorus and Elemental Phosporus | 1991 |
| 16. | Non-Biodegradable Detergents and Other Surface-Active Substances | 1991 |
| 17. | Thermal Discharges | 1991 |
| 18. | Acid or Alkaline Compounds | 1992 |
| 19. | Substances having adverse effects on the Oxygen content | 1992 |
| 20. | Barium, Uranium and Cobalt | 1992 |
| 21. | Cyanides and Fluorides | 1993 |
| 22. | Substances, of a non-toxic nature, which may become harmful owing to the quantities discharged | 1993 |
| 23. | Organosilicon Compounds | 1993 |
| 24. | Antimony, Tin and Vanadium | 1994 |
| 25. | Substances which have a deleterious effect on the taste and/or smell of products for human consumption | 1994 |
| 26. | Biocides and their derivatives not covered in annex I | 1994 |
| 27. | Titanium, Boron and Silver | 1995 |
| 28. | Molybdenum, Beryllium, Thallium and Tellurium | 1995 |

## REFERENCES

1. Miller, A., The Mediterranean Sea, A. Physical aspects. In: B.H. Ketchum (Ed.), Estuaries and Enclosed Seas, Elsevier, pp. 219-238, 1983.

2. UNEP, Mediterranean Action Plan and the Final Act of the Conference of Plenipotentiairies of the coastal States of the Mediterranean Region for the Protection of the Mediterranean Sea. United Nations, N.Y., 1978.

3. UNEP, The Co-ordinated Pollution Monitoring and Research Programme (MED POL - PHASE I); Programme Description. Regional Seas Reports and Studies No. 23, Rev. 1. UNEP, Athens, 1983.

4. UNEP/IAEA/IOC, Reference Methods and Materials: A programme of support for regional and global marine pollution assessments. UNEP, Athens, 1990.

5. UNEP, State of the Mediterranean Marine Environment. MAP Technical Reports Series No. 28. UNEP, Athens, 1989.

6. Bernhard, M. and A. Renzoni, Mercury concentration in Mediterranean marine organisms and their environment: natural or anthropogenic origin. Thalassia Jugosl., 13:265-300, 1977.

7. Sara, R., Sulla biologia dei tonni (Thunnus thynnus L.), modelli di migrazione ed osservazioni sui meccanismi di migrazione e di comportamento, Boll. Pesca Piscic. Idrobiol.: pp. 217-243, 1973.

8. Thibaud, Y., Pollution par les métaux lourds en Méditerranée. Etude chez les poissons des mécanismes de contamination et de décontamination. Convention no. 75-08, Inst. Scientif. et Technique des Pêches Maritimes, Nantes, 1979.

9. Buffoni, G., M. Bernhard and A. Renzoni, Mercury in Mediterranean tuna. Why is their level higher than in Atlantic tuna? A model. Thalassia Jugosl. 18:231-243, 1982.

10. Bernhard, M., Mercury accumulation in a pelagic foodchain. In: Environmental Inorganic Chemistry, Matertell, A. E. and K. J. Irgolic, (Eds.) Weinhein: Verlag Chemie. pp. 349-358, 1985.

11. Cerrati, G., Determinazione del mercurio oganico ed inorganico nel muscolo del fegato e nel cervello di organismi appartenenti ad une catena alimentare pelagica tramite spettrofotometria ad assorbimento atomico con fornetto di grafite e speciazione del mercurio organico per via gas-chromatografica. Thesis, Univ. Pisa, 1987.

12. Astier-Dumas, M. and G. Cumont, Consommation hebdomadaire de poisson et teneur du sang et des cheveux en mercure en France. Ann. Hyg. L. Fr. - Med. et Nut. 11:135, 1975.

13. Paccagnella, B., L. Prati and A. Bigoni, Studio epidemiologico sul mercurio nei pesci e la salute umana in un'isola Italiana del Mediterraneo. L'Igiene Moderna 66:479-503, 1973.

14. WHO, Environmental Health Criteria 1. Mercury. World Health Organization, Geneva, 1976.

15. Riolfatti, M., Ulteriori indagini epidemiologiche sulle concentrazioni di mercurio nel pesce alimentare e nel sangue e capelli umani. L'Igiene Moderna 70:169-185, 1977.

16.  Bacci, E., G. Angotzi, A. Bralia, L. Lampariello and E. Zanette, Etude sur une population humaine exposée au methylmercure par la consommation de poisson. *Rev. Inter. Oceanogr. Medicale* 41/42:127-141, 1976.

17.  Nauen, C., G. Tomassi, G. P. Santaroni and N. Josupeit, Results of the first pilot study on the chance of Italian seafood consumers exceeding their individual allowable daily mercury intake. *Journ.Etud.Pollut.CIESM.*, 6:571-583, 1983.

18.  Huynh-Ngoc, L. and R. Fukai, Levels of trace metals in open Mediterranean surface waters - a summary report. *Journ.Etud.Pollut.CIESM.*, 4:171-175, 1979.

19.  Kremling, K. and H. Petersen, The distribution of zinc, cadmium, copper, manganese and iron in waters of the open Mediterranean Sea. Berlin, Gebrueder Borntraeger. "Meteor" Forsch. Ergebnisse, Reihe A/3 (23):5-14, 1981.

20.  UNEP, Report on the state of pollution of the Mediterranean Sea (UNEP/IG.56/Inf.4). Athens, 1985.

21.  UNEP/FAO, Assessment of the present state of pollution by cadmium, copper, zinc and lead in the Mediterranean Sea (UNEP/WG.144/11). Athens, 1986.

22.  Copin-Montegut, G., P. Courau and E. Nicolas, Distribution et transferts d'éléments traces en Mediterranée occidentale. Nouveaux resultats Phycomed. *Journ.Etud.Pollut.CIESM.*, 7:111-117, 1985.

23.  Stoeppler, M., C. Mohl and H. W. Nuernberg, Total Arsenic in sea water and marine organisms: a comparative study from the Mediterranean Sea and selected regions of the oceans. *Journ.Etud.Pollut.CIESM.*, 5:281-283, 1981.

24.  WHO, Environmental Health No. 8: Industrial Wastewater in the Mediterranean Area. WHO, Copenhagen, 1986.

25.  Whitehead, N. E., B. Oregioni and R. Fukai, Background levels of trace metals in Mediterranean sediments. *Journ.Etud.Pollut.CIESM.*, 7:233-240, 1985.

26.  Bernhard, M., Levels of trace metals in the Mediterranean. *Journ.Etud.Pollut.CIESM.*, 6:237-243, 1985.

27.  ICES, Final Report of a Working Group for the international study of the pollution of the North Sea and its effects on living resources and their exploitation. Coop. Res. Rep. ICES. no. 39, 1974.

28.  ICES, The ICES co-ordinated monitoring programmes 1976 and 1977. Coop. Res. Rep. ICES no. 72, 1977.

29.  ICES, The ICES Co-ordinated monitoring programme in the North Sea, Coop. Res. Rep. ICES. no. 58, 1977.

30.  Risebrough, R. W., B. W. De Lappe and T. T. Schmidt, Bioaccumulation factors of chlorinated hydrocarbons between mussels and seawater. *Mar. Pollut. Bull.* 7:225-228, 1976.

31.  Picer, N. and M. Picer, Monitoring of chlorinated hydrocarbons in water and sediments of the North Adriatic coastal waters. *Journ.Etud.Pollut.CIESM.*, 4:133-136, 1979.

32.  Fowler, S. W., PCBs and the Environment. The Mediterranean marine ecosystem, In: PCBs and the environment, J.S. Waid (Ed.) Vol.3, Ch.8, CRC Press Inc., Boca Raton, Fl., pp. 209-239, 1986.

33. UNEP/IMO/IOC, Assessment of the present State of Pollution by Petroleum Hydrocarbons in the Mediterranean Sea (UNEP/WG.160/11). Athens, 1987.

34. UNEP/WHO, Evaluation of MED POL - Phase II Monitoring data. Part II - Micro-organisms in coastal waters. UNEP(OCA)/MED WG. 5/Inf. 4. Athens, 1989.

35. UNEP/UNESCO/FAO, Eutrophication in the Mediterranean Sea: Receiving capacity and monitoring of long-term effects. MAP Technical Reports Series No. 21. UNEP, Athens, 1988.

36. UNESCO, Eutrophication in coastal marine areas and lagoons: a case study of "Lac de Tunis". UNESCO reports in Marine Science no. 29. Paris, 1984.

37. Stirn, J., A. Avcin, J. Cencelj, M. Dorer, S. Gomiscek, S. Kveder, A. Malej, D. Meischner, I. Nozina, J. Paul and P. Tusnik, Pollution problems of the Adriatic Sea. An interdisciplinary approach. Rev. Intern. Oceanogr. Méd., 35-36:21-78, 1974.

38. Stirn, J., Eutrophication in the Mediterranean Sea. In: UNEP/UNESCO/FAO:Eutrophication in the Mediterranean Sea: Receiving capacity and monitoring of long-term effects. MAP Technical Reports Series, No. 21, UNEP, Athens, 1988.

39. Marchetti, R., G.F. Gaggino and A. Provini, Red tides in the Northwest Adriatic. In: UNEP/UNESCO/FAO; Eutrophication in the Mediterranean Sea: Receiving capacity and monitoring of long-term effects. MAP Technical Reports Series No. 21. UNEP, Athens, 1988.

40. Pucher-Petkovich, T. and I. Marasovic, Evolution de quelques populations diatomiques dans une aire soumise à l'eutrophisation (Adriatique Centrale). Rapp. Comm. Internat. Mer Médit. 26., 1978.

41. Yannopoulos, C., The annual regeneration of the Elefsis Bay zooplankton ecosystem, Saronikos Gulf. Rapp. Comm. int. Mer Médit. 23 (9):109-111, 1976.

42. Zarkanella, A.J., The effects of pollution-induced oxygen deficiency on the benthos in Elefsis Bay, Greece. Marine Environ. Res., 1979, pp. 191-207, 1979.

43. Fukai, R., S. Ballestra, M. Thein and J. Guion, Input of transuranic elements through rivers into the Mediterranean Sea. Impacts of Radionuclide Releases into the Marine Environment. IAEA, Vienna, 1981.

44. Ballestra, S., E. Holm, A. Walton and N.E. Whitehead, Fallout deposition at Monaco following the Chernobyl accident. J. Environ. Radioactivity, 5, pp. 391-400, 1987.

45. UNEP/WHO, Assessment of the present state of microbial pollution in the Mediterranean Sea and proposed control measures (UNEP/WG.118/6). UNEP, Athens, 1985.

46. UNEP/FAO/WHO, Assessment of the state of pollution of the Mediterranean Sea by mercury and mercury compounds. MAP Technical Reports Series No. 18, UNEP, Athens, 1987.

47. UNEP/WHO, Assessment of the state of microbial pollution of shellfish waters in the Mediterranean Sea and proposed measures (UNEP/WG.160/10) UNEP, Athens, 1987.

48. Assessment of the situation regarding used lubricating oils in the Mediterranean basin and suggested progressive measures for their elimination as marine pollutants (UNEP(OCA)/MED WG.3/Inf.4). UNEP, Athens, 1989.

49.  UNEP/FAO/WHO,  Assessment of the state of pollution of the
     Mediterranean Sea by cadmium and cadmium compounds.  MAP Technical
     Reports Series No. 34.  UNEP, Athens, 1989.

50.  UNEP/FAO/WHO/IAEA,  Assessment of organotin compounds as marine
     pollutants in the Mediterranean.  MAP Technical Reports Series No.
     33.  UNEP, Athens, 1989.

51.  UNEP/FAO/WHO/IAEA,  Assessment of the state of pollution of the
     Mediterranean Sea by organohalogen compounds (UNEP(OCA)/MED
     WG.3/Inf.6).  UNEP, Athens, 1989.

52.  UNEP/FAO/WHO/IAEA,  Assessment of the state of pollution of the
     Mediterranean Sea by organophosphorus compounds (UNEP(OCA)/MED WG.
     25/Inf.4).  UNEP, Athens, 1991.

53.  UNEP/FAO/IOC,  Assessment of the state of pollution of the
     Mediterranean Sea by persistent synthetic materials which may
     float, sink or remain in suspension (UNEP(OCA)/MED WG. 25/Inf.5).
     UNEP, Athens, 1991.

54.  UNEP/IAEA,  Assessment of the state of pollution in the
     Mediterranean Sea by radioactive substances (UNEP(OCA)/MED WG.
     25/Inf.8).  UNEP, Athens, 1991.

55.  UNEP/WHO,  Assessment of the state of pollution of the
     Mediterranean Sea by pathogenic microorganisms (UNEP(OCA)/MED WG.
     25/Inf.7).  UNEP, Athens, 1991.

56.  UNEP,  Common Measures adopted by the Contracting Parties to the
     Convention for the Protection of the Mediterranean Sea against
     Pollution.  MAP Technical Reports Series No. 38.  UNEP, Athens,
     1990.

57.  UNEP,  Reports of the Seventh Ordinary Meeting of the Contracting
     Parties to the Convention for the Protection of the Mediterranean
     Sea against pollution and its related protocols (UNEP(OCA)/MED IG.
     2/4).  UNEP, Athens, 1991.

# PREDICTED SEA-LEVEL RISE IN THE WIDER CARIBBEAN: LIKELY CONSEQUENCES AND RESPONSE OPTIONS

LEONARD A. NURSE
Project Manager
Coastal Conservation Project Unit
Oistins, Christ Church, Barbados, W.I.

## ABSTRACT

The coastal and marine resources of the Wider Caribbean constitute some of the region's most valuable assets. If predicted rates of sea-level rise were to materialize, some of these resources could be significantly impacted. This paper assesses the importance of the region's coastal resources, and suggests various criteria for identifying those areas and resources at greatest risk. Some of the more significant physical, ecological and socio-economic consequences of predicted sea level rise are reviewed and evaluated. Response options are examined in the context of 'local' constraints, while some suggestions are offered for improving the region's ability to cope with future sea-level rise.

## 1.0 THE COASTS OF THE WIDER CARIBBEAN: THEIR SIGNIFICANCE IN NATIONAL DEVELOPMENT

The Wider Caribbean is defined by the United Nations Environment Programme as that area encompassing 'the marine environment of the Gulf of Mexico, the Caribbean Sea and the areas of the Atlantic Ocean adjacent thereto, south of 30° north latitude and within 200 nautical miles of the Atlantic coasts of the states referred to in article 25 of the Convention' [1]. See also figure 1.

In the Caribbean, as elsewhere, the coastal zone has played an important role in the social and economic life of the region, ever since settlement. For many states the coastal zone is now the primary focus of most activities, with most major urban centres either located along the littoral itself or a few kilometres inland. In the Eastern Caribbean, for example, over 50% of the

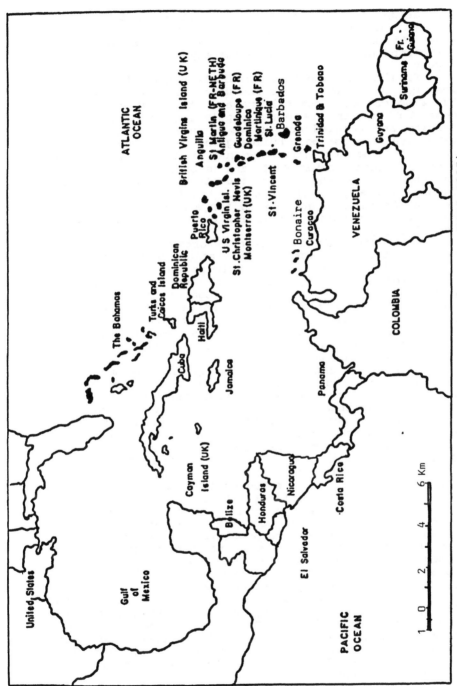

Figure 1. The Wider Caribbean - (Source: UNEP 1987)

population resides within two kilometres of the coast. For the entire Wider Caribbean region, a figure of at least 40% would be a realistic estimate. In Barbados the corresponding figure is approximately 60% [2]. See also figure 2.

Fishing, an important pursuit throughout the region, does not only exploit the resources of the marine zone but also consumes coastal space. It is estimated that some 6-8% of the region's population is involved in fishing and related activities. In addition, coastal sites are used for boat building and repair and serve as traditional market places for much of the catch. Apart from providing employment and supplying protein to Caribbean diets, the fishing industry is a net earner of foreign exchange. The Central American countries of Belize, Costa Rica and Mexico for example export large quantities of shrimp, lobster, conch and frozen fish annually [3].

Agriculture in its various forms is important on many coasts. The scale of operation varies from one location to the next, but includes small subsistence plots for food crop production and large estates producing rice (e.g. along Guyana Coastal Plain), sugar and other commercial crops. In spite of the concerted industrial thrusts which began in the mid to late 1960's, agriculture is still important to the region, employing significant numbers and contributing more than 10% of the region's Gross Domestic Product [4].

Tourism infrastructure now occupies large amounts of coastal space. Hotels, guests houses, condominiums and ancillary facilities line many segments of the coasts of the Wider Caribbean. This is as true of the northern portion of the region (Florida and Gulf of Mexico area) as it is of the small island states to the south. Tourism assumes particular prominence in the Eastern Caribbean islands, where it generates significant employment and valuable foreign exchange for these small, open economies. Visitor spending as a percentage of GDP is high for most countries (Table 1).

Industry too has found its way to the coast, on account of some obvious locational advantages. These advantages include (a) the proximity to port facilities for import of raw materials and export of finished products, as in the case of oil refining and manufactured goods and (b) access to large volumes of 'cooling water', in the case of power generation plants.

The coast is used for many other activities including forestry, military installations, leisure and recreational pursuits. The consequences of global climate change, particularly sea-level rise, therefore raises profound concerns in a region where the littoral and marine zone play such a pivotal role in social and economic development.

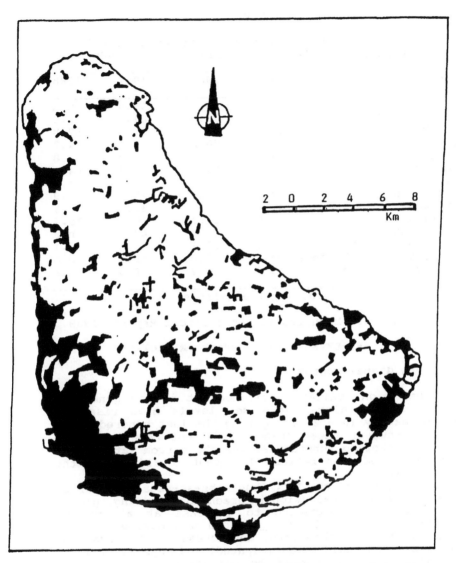

Figure 2. Barbados distribution of settlement. Note the heavy concentration along the west, south and south-east coasts.

(Source: Barbados Physical Development Plan, 1988)

TABLE 1

Visitor spending as a percentage of Gross Domestic
Product for selected countries, 1985 - 1988

| Country | 1985 | 1986 | 1987 | 1988 |
|---|---|---|---|---|
| Anguilla | 101.9 | 96.1 | 106.8 | n.a |
| Antigua & Barbuda | 77.0 | 78.5 | 81.8 | 77.8 |
| Bahamas | 53.3 | 52.6 | 51.1 | 52.8 |
| Barbados | 28.1 | 28.2 | 30.0 | 34.2 |
| Belize | 6.3 | 22.8 | 23.2 | 21.8 |
| Bermuda | 74.7 | 83.0 | 87.7 | 84.9 |
| B.V.I. | 121.0 | 138.6 | 143.3 | n.a |
| Cayman Islands | 27.9 | 27.3 | 28.6 | 31.9 |
| Grenada | 33.6 | 38.3 | 37.7 | n.a |
| Jamaica | 20.0 | 21.0 | 20.4 | 15.7 |
| St. Lucia | 38.7 | 46.5 | 47.5 | n.a |
| St. Vincent & Grenad. | 24.8 | 28.1 | 30.3 | 35.1 |
| Turks & Caicos Islands | 34.7 | 57.0 | 58.1 | n.a |
| U.S. Virgin Islands | 49.9 | 50.2 | 50.0 | n.a |

n.a. = not available

Source: Caribbean Tourism Statistical Report, 1989

## 2.0      SEA-LEVEL RISE: CURRENT PREDICTIONS

### 2.1  The Global Context

Based on available evidence, it is now generally
accepted that global mean sea levels have risen by
approximately 10 - 20 cm. during the last hundred years
[5].    However these changes have not been spatially
uniform, showing substantial variation from latitude to
latitude, and even between different localities at the
same latitude.

According to the IPCC 'business-as-usual' scenario,
i.e. assuming no reduction in greenhouse forcing, global
sea levels are expected to increase at a rate of
approximately 6 cm/decade ($\pm$ 3 - 10 cm), during the next
century [5]. At this rate, an increase of the order of
20 cm by 2030 could be anticipated, and approximately 65
cm by 2100.    It should be emphasized that even if
greenhouse gas emissions were stabilized as early as
2030, sea level would still continue to increase for more
than an additional 20 cm by 2100 [6].  See also figures
3(a) and (b).  Other predictions are more pessimistic,
some suggesting that there could be a rise of 100 m well
before 2100 [7, 8].    However most current research is
based on the WMO/UNEP-IPCC model now widely regarded as
the   most   plausible,   in   spite   of   its   recognized
limitations.

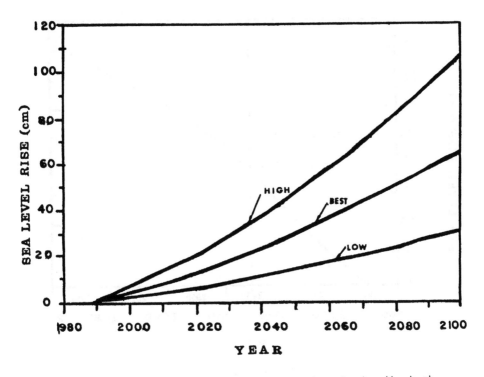

Figure 3(a). IPCC 'Business-as-Usual' Scenario, showing the best estimate (20 cm by 2030, 65 cm by 2100) and high and low estimates.

Source: WMO/UNEP-IPCC, 1990

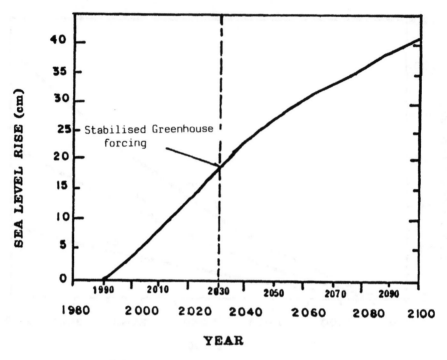

Figure 3 (b). Graph illustrating predicted sea-level rise under the scenario of stabilised greenhouse forcing 2030. Note that the magnitude of increase for the period 2030 - 2100 is similar to the prediction for the years 1990 - 2030.

Source: WMO/UNEP-IPCC, 1990

## 2.2   The Record for the Wider Caribbean

While the long-period sea level record for the Wider Caribbean is inadequate, the best available evidence suggests that mean sea level has been increasing at a rate of about 0.36 cm/yr, with values ranging from 0.22 cm/yr at Key West, Florida to +1.0 cm/yr along the Texas coast [9].   But a fall in relative sea-level has also been recorded at some sites, notably along the Mexico coast, where a rate of -0.3 cm/yr is estimated  [9].

Relying on these and other similar records, the UNEP Task Team of Experts for the Wider Caribbean concludes that a mean sea-level rise of approximately 10 cm by the year 2025 might be realistic for the region [9].   This value should however be accepted with some caution.   In the first place, there is significant spatial variation in the record;   secondly,   some   of   the   data   covers relatively short periods; and thirdly, the existing data base is not spatially uniform, with areas such as the Lesser Antilles having sparse coverage.

Considering the limitations of the Caribbean record, the ensuing discussion is therefore based on the most recent IPCC estimate of 20 cm by 2030.

## 3.0        VULNERABILITY TO SEA-LEVEL RISE: DEFINING THE CRITERIA FOR THE WIDER CARIBBEAN

3.1  The threat of sea level rise of any given magnitude inevitably raises the issue of 'vulnerability'.   Given the   variation   in   physical   character   of   the   Wider Caribbean, levels of vulnerability will vary both between and within countries.   It is generally agreed that the extent to which a given location is vulnerable to sea level   rise,   varies   according   to   such   factors   as geological and geomorphological history and character, elevation, ecology and other 'physical' considerations [10].   However while this author supports that general view,   a case is being made for vulnerability to be defined in more 'human' terms.   Indeed it can be argued that sea level rise is considered a real threat only on account of

(i)    the level of human presence (permanent and transient) associated with the coast and

(ii)   the significance or perceived value of the natural and man-made systems ('resources') likely to be impacted.

Thus the criteria used here to define vulnerability will therefore reflect both physical and human factors. No attempt is necessarily made to assign a rank to any group.   For the criteria simply seek to identify those areas which would appear to be at greatest risk.

i.    Poorly drained low-lying coastal plains at or below sea level with dense populations and which have significant economic activity and infrastructure along the littoral: e.g. Guyana (Georgetown is 1 m below mean sea level), the coasts of Suriname and Venezuela, segments of the Mississippi and Florida coasts, Belize.

ii.    Areas which may be experiencing subsidence and simultaneous sea-level rise: e.g. Port-au-Prince (Haiti), Puerto Cortes (Honduras) and parts of the Texas - Louisiana coast [9].

iii.    Small island states which are heavily coastal-dependent (for instance tourism) and which owing to limitations of size, have no real hinterland. In other words those islands which can legitimately be described as completely 'coastal': e.g. Archipelago of the Lesser Antilles, Bahamas, Turks and Caicos, British Virgin Islands.

iv.    Coasts with critical habitats and ecosystems (coral reefs, wetlands, seagrass beds etc.) which are likely to be impacted, irrespective of whether the associated resources are presently being exploited or not: e.g. Insular Caribbean, parts of Central America such as Belize Barrier Reef Zone.

v.    Areas which are already severely stressed from other influences mainly anthropogenic e.g. pollution and non-enforcement of coastal building codes: Segments such as Kingston Harbour area and the built-up coasts of the islands of the Lesser Antilles, fall into this class.

vi.    The region's poorest countries (e.g. Haiti) which are presently grappling with basic social and economic problems (housing, health, unemployment, political instability etc). Such countries are far less able to divert scarce technical and financial resources to address 'other issues', than better-off states. Countries falling within this group must be considered quite vulnerable, as they could find themselves completely at the mercy of donor countries and agencies. Under a scenario of accelerated global sea-level rise, heightened competition for external assistance must be anticipated.

It should be emphasized that the above classes are not intended to be mutually exclusive, since there is a range of countries which can be identified as belonging to more than one group.

## 4.0 EFFECTS OF SEA-LEVEL RISE ON WIDER CARIBBEAN

There can be little doubt that if predicted rates of sea level rise for the Caribbean were to materialize, the economic and social effects could be widespread and far-reaching. It has already been demonstrated elsewhere in this paper, that the coastal and marine resource base is a critical part of the region's life-support system. Thus any dislocation of the resource base could have disastrous consequences for the region.

### 4.1 Physical, Ecological and Environmental Impacts

4.1.1    Coastal erosion:    While the measured rate of sea-level rise for the Wider Caribbean may be low in comparison with estimates from elsewhere, a rise of 20 cm would have disproportionately great effects on the shoreline.  Shoreline retreat will inevitably occur as the sea encroaches and as wave energy increases in the nearshore zone, consequent upon increasing water depth and greater wave heights. Many of the region's coasts and beaches are already experiencing rapid erosion, as a result of other influences. Among these are degradation of coral reefs, nutrient enrichment of nearshore waters, construction in the active beach zone (which impacts on beach equilibrium), sand mining and poor sediment management in general [11]. Relative sea level rise of even a limited magnitude could exacerbate the current erosion trend in many countries.

4.1.2    Inundation and Flooding:    Coastal floods, already widespread especially along low-lying coastal plains (e.g Guyana; Barbados west and south coasts; Puerto Cortes, Honduras; Port-au-Prince, Haiti and subsiding sections of the Texas and Louisiana coasts) would increase as a result of rising sea levels [9, 12, 13].  Given the geomorphological and tectonic variations along Caribbean coasts, the actual magnitude of flood events cannot be easily predicted from generalized models.  Thus detailed case studies of vulnerable low-lying coastal zones must be undertaken if the problem is to be properly understood.

Yet it would be safe to predict that where extreme

events such as storms and hurricanes occur, the accompanying storm surge would most likely become exaggerated. And if these events were to coincide with astronomical high tide, consequential inundation might extend inland, way beyond the immediate coastal fringe. It is projected for example that if a fully developed hurricane were to make landfall at Barbados at high tide, 2-metre waves would reach some 80 metres inland on segments of the low-lying south coast of the island [14]. If a sea level rise of even 10 cm above datum were factored into the equation, wave run-up, erosion and flooding would have disastrous consequences on the dense ribbon of settlement that exists. The seriousness of such a threat is not confined to Barbados; for most of the other low-lying areas such as the Bahamas, Bermuda, Belize city, the Mississippi Delta region and the Florida Keys [15] would also be at risk.

4.1.3    <u>Saline intrusion:</u>    There is general agreement that global sea level rise may lead to saline intrusion into coastal aquifers [16, 17, 18].   In many states especially the low coral islands, there is almost total dependence on groundwater for domestic, industrial and irrigation purposes.  The threat of saline intrusion in these countries becomes more real when it is considered that groundwater supplies are already under tremendous pressure.  Where drawdown is not compensated by adequate recharge, the fresh water lens shrinks thereby becoming more susceptible to saline intrusion [19].

Increasing salinisation would also occur directly from the flooding of deltas, lagoons, marshes, wetlands and other low-lying coastal lands.  This phenomenon is already being observed in areas of the world such as the Maldives[20, 21].   Vulnerable deltas and estuaries including the Orinoco, Rio Grande, northern Gulf of Mexico, and the Guyana Coastal Plains would be at increased risk.  Thus the various activities including agriculture, which are currently based on such areas, might assume a lesser economic significance with accelerated sea-level rise.

4.1.4    <u>Coral reefs</u>:    The western Caribbean Sea is fringed by the world's second largest coral reef system. The extent to which reefs of the Wider Caribbean region can cope with increasing sea levels is influenced by many factors: reef type, bathymetry, reef depth, sedimentation rates and a variety of anthropogenic influences including pollution.  For the region as a whole, the average rate of growth for most reefs is between 1 - 20 cm/yr.  While not all reefs accumulate at these rates, it is known that most corals could keep pace with a rise of even 20 cm by

2030,[1] especially species like the branching _Acropora palmata_ and _Acropora cervicornis_ [22,23,24,25,26].

It must be cautioned that the above projections assume no acceleration in rate of sea level rise. Available data from the Caribbean and elsewhere suggests that even those reefs dominated by the branching _Acropora_ which accumulate at between 7 - 8mm/yr could not easily survive a rise in excess of 10 mm yr [22,25]. If greatly increased rates of sea level rise were to occur, then many coral reef communities might find it difficult to survive. This would have a dislocating effect on coastal stability and marine resources in the Wider Caribbean. For the reefs provide a nursery and habitat for many species of fish which form an essential part of the diet of the peoples of the Wider Caribbean. Almost 200 of the 350 known reef-associated species are harvested commercially [28]. Lobsters (especially _Panulirus argus_) and conch (_Strombus gigas_) are exported in significant quantities [29]. At the same time coral reefs are vital sources of sediment for the beach sand budget along many Caribbean coasts. Moreover they absorb, reflect and dissipate incoming wave energy and thus reduce beach erosion at may sites [11].

Perhaps the greatest challenge to Caribbean coral reefs is not sea-level rise _per se_, but rather the ability of these systems to cope with the rise, in a situation of elevated sea surface temperature. Most corals have a narrow range of tolerance to temperature changes. Optimum water temperatures for coral growth are between 20 - 28°C and quantitatively small variations from this range can have a negative impact on growth rates. If sea surface temperatures in the Wider Caribbean were to increase by 1.5°C (i.e. WMO/ICSU/UNEP 1985 baseline scenario), then a sea-level rise of 20 cm by 2030 could be problematic. Sudden increases in water temperature are known to cause 'bleaching' or the loss of symbiotic zooxanthellae [24,28]. This phenomenon frequently leads to increased morbidity and mortality of reefs. The recent bleaching events of 1982 - 83 and 1987 occurring in the Caribbean have been attributed to increasing ocean temperatures triggered by El Nino - Southern Oscillation. The effects were particularly marked in the Pacific, where by 1983, 70 - 95% of the live corals of eastern Pacific Reefs (to depths of 15 - 18 metres) had died, with some species becoming extinct [30]. Any combination of rising sea level, elevated water temperature and anthropogenic influences would therefore be disastrous for Caribbean reefs.

---

[1]. This assumes that other factors are insignificant.

4.1.5   __Mangroves__:   Mangroves are unique communities of sheltered tropical shorelines and have great ecological and economic value. They reduce wave energy, provide habitats for rare species and function as spawning grounds for shell and fish species.   Any significant threat to the survival of these ecosystems resulting from predicted climate change and sea level rise events, must therefore be of great concern to regions like the Wider Caribbean.

It is difficult to generalize about the response of mangrove communities to anticipated sea-level rise, as there is variation in the behaviour of different species to changing habitat parameters [22]. Limited research also suggests that _Rhizophora_ might have greater resilience to sea level rise than _Avicennia_, particularly because of the height of their aerial roots. Nevertheless there is now growing evidence to suggest that most species could cope with predicted sea level change, i.e. 10 - 20 cm by 2030, provided the communities are not limited by steep coastlines[2] [22].   Mangrove-derived peat accumulation during the Holocene at rates of up to 8.8 cm/100 yrs in Grand Cayman [31], and southeast Florida [30] support this view.

It would appear that mangroves can adjust to rising sea-levels of the order of around 8 cm/100 yrs, but show evidence of stress in coping with levels between 8 and 12 cm/100 yrs [32].   As in the case of other coastal ecosystems, the greatest uncertainties affecting their ability to cope are (a) whether or not there is an increase in the predicted rate of sea level rise (b) the effect of elevated temperature as a stress factor and (c) the extent to which anthropogenic forces influence habitat conditions.

4.1.6   __Seagrass beds:__   It is not expected that sea level rise of 20 cm by 2030 should cause significant dislocation to the region's six most common seagrasses. Seagrass beds and micro-algae are believed to be capable of shoreward migration, once the submerged bottom provides a favourable habitat for spores and seedlings [33].   Climate change may however be disruptive to the seagrass beds on account of changes in other parameters, such as temperature and precipitation, which may exceed the tolerance of some species.   In addition moderate impact may occur if sea level rise is accompanied by factors such as change in light quality, effect of herbivores or changes in wave energy or bottom slope [9]. For practical purposes, the latter conditions could be considered as the 'worse case scenario'. But generally

---

2.   It must be assumed that such physical constraint would restrict the capacity of the mangrove communities to migrate landward.

there appears to be concurrence in the view that seagrasses have the capacity to withstand currently predicted rates of sea level rise.

## 4.2. Economic And Social Impacts

4.2.1    Agriculture:    repeated flooding of low-lying lands which may accompany sea-level rise could raise the salinity level of coastal soils.  Under such circumstances crops which have a low tolerance to salt could be displaced.  It is believed for example that the extensive cultivation now practiced along the Guyana coast might be heavily impacted.  The vulnerability of agricultural resources to the threat of sea-level rise will obviously vary according to elevation.  Clearly the most vulnerable zones will be those agricultural lands presently at or below mean sea-level.

4.2.2    Fisheries:    A sea level rise of 10 - 20 cm by 2030 in itself may not be seriously disruptive to fisheries resources in the Wider Caribbean.  However, if increased turbidity accompanies the rise in sea level, dislocation would result.  It should also be pointed out that coastal wetlands, estuaries and deltas form prime spawning and nursery grounds for a variety of shellfish, crabs, shrimps and fishes.  Coastal erosion occasioned by sea level rise could impact on local nearshore currents, disrupting spawning and breeding regimes.  Sea level rise may also affect the migratory patterns of commercially important species [33].

It has been suggested that salinity changes as a possible consequence of sea level rise, would adversely affect estuarine - dependent species in some areas, including the Mississippi and Orinoco deltas, the Florida Everglades and along the Guyana Coasts [9].  There is also the possibility that some juveniles could be placed under severe stress where conditions of hyper-salinity exists. This additional stress factor  would increase if a significant elevation of sea surface temperature occurs simultaneously.

4.2.3    Tourism:    Over 95% of all tourism activities in the Wider Caribbean is marine and coastal oriented. The North American Demand Study for Caribbean Tourism [32] has shown that visitors identify high quality beaches as one of the most important factors influencing their choice of the Caribbean, as a holiday destination. The erosion of prime beachlands resulting from sea level rise could therefore affect the region's ability to attract large numbers of visitors.  Ironically, much of the region's most expensive tourism infrastructure is located along some of the most vulnerable coasts.  The islands of the Eastern Caribbean, Jamaica, the Bahamas, Bermuda and the Florida coast fall into this category.

On the positive side, shoreline migration would almost certainly lead to the formation of 'new' beach areas, which themselves could become potential tourism centres.

Reef-based activities such as scuba diving, snorkeling and recreational fishing, are important 'pull factors' in Caribbean tourism. Thus, in the event that the coral reefs suffer adversely from rising sea levels, the industry could be jeopardized. Tourism is an industry which responds quickly to perceived changes in the quality of the product being offered, whether real or imaginary. The swiftest response is to look to other destinations. The Wider Caribbean, especially the small islands, are so dependent on tourism, that such an eventuality would be completely disastrous for many economies. In the British Virgin islands earnings from marine tourism alone are estimated to be around US $14 million annually [28]. The tourism contribution to Gross Domestic Product in Antigua is approximately 77%, while in the Bahamas and Barbados the figure is about 50% and 25% respectively. In the smaller, poorer countries of the region where resources are already scarce, considerable strain would therefore be placed on other sectors of the economy.

4.2.4    Water resources:    The possibility of saline intrusion into the region's aquifers has already been addressed in section 4.1.3. The quality and quantity of available water will impact on all areas of economic and social development. The only real option for the region is the immediate implementation of efficient water resources management and conservation strategies. With current supplies already under threat from rapid drawdown, any additional stress imposed by the consequences of sea level rise will set back the region's development process in a significant way.

4.2.5    Infrastructure:    coastal infrastructure such as sea and airports, roads and coastal protection structures, would clearly be at greater risk with a sea level rise of 20 cm. The combined effects of flooding and increased expenditure of wave energy, especially in low-lying areas, would pose serious problems to infrastructure. The settlements along the Belize and Guyana coasts for instance, would be at great risk [22]. The situation would worsen if there is an increase in the frequency and magnitude of severe events, such as storm surges.

The location of major power plants in the Caribbean Basin is also a major concern. Nearly all countries have their facilities located on the coast, in order to take advantage of abundant supplies of cooling water. It has been demonstrated that some power plants could be crippled, if sea level rise exceeds cooling-water intake

and other design specifications [33]. Most of the smaller
economies depend on a single power plant, and down-time
during 'normal' operational conditions is severely
disruptive. The design and location of planned and
existing structures will have to be critically reviewed,
in order to mitigate the additional impact of predicted
sea level rise.

There has been a proliferation of coastal protection
structures in the Caribbean during the last thirty years
or so. This trend has accompanied the rapid expansion of
coastal settlements, industrial plants and heavy
investments in tourism plant. It is possible that with
a 20 m rise in sea level, additional structures may
become necessary, while existing ones may have to be
redesigned. Where new structures are being contemplated,
careful consideration should be given to the possible
consequences of sea level rise, so that appropriate
design criteria might be applied.

4.2.6     Resettlement:          The possibility of massive
relocation of vulnerable coastal settlements remains one
of the most daunting prospects for the Wider Caribbean.
Many settlements are presently at or about sea level
(e.g. Barbados, Bahamas, Turks and Caicos, Port-au-
Prince, Gulf of Honduras), while others such as those
along the Guyana Coast, are below mean sea level. When
the combined effects of increased flooding, ground water
contamination, coastal resources loss, and
infrastructural damage are contemplated, some
resettlement of coastal populations seems almost
inevitable. This will not only be socially dislocating
for the peoples themselves, but it would also require an
injection of funds which few countries in the region can
afford.

The implications of resettlement are even more
far-reaching for the island states of the Wider
Caribbean. Some countries are so small and are already
so densely populated, that resettlement within their own
national boundaries may be physically impossible.

4.2.7     Human Health: The combined effects of elevated
temperatures and rising water levels can lead to an
increased incidence of certain tropical diseases [33].
A greater frequency of flooding and higher water levels
would provide favourable environments for the
proliferation of many water-borne bacteria which carry
disease [33]. It is postulated that health problems
occasioned by climate change and sea level rise, could
possibly affect all the productive sectors of Caribbean
economies, due to health-related absenteeism [9]. Under
these circumstances worker productivity and overall
economic output would decline.

Overall, the general expectation is that the region's health care systems should be able to cope with climate-related health problems at least up to the year 2030. This view must nevertheless be considered with some caution, since there are many variables which could determine the region's ability to adapt. Increased man-induced stresses from pollution, poor sanitation, solid waste disposal practices and other environmental risk factors cannot be ignored.

## 5.0 RESPONSE OPTIONS FOR WIDER CARIBBEAN

### 5.1 Available Options

There is now a global consensus that adaptive strategies to sea level rise can be classified into three broad categories: retreat, protection and accommodation [13,36,37]. The retreat option implies a relocation of human activities away from threatened coasts, allowing ecosystems to migrate shoreward as they adjust to the new conditions. Accommodation, like retreat, requires no measures to prevent the encroachment of the sea; but unlike retreat, there is no shift away from or abandonment of the coast. Measures such as the raising of buildings, the cultivation of halophytes and the conversion of agricultural activities to aquaculture would be associated with accommodation. Protection involves the applications of 'hard' (seawalls, revetments etc.) and 'soft' (beach nourishment, revegetation etc) engineering options to prevent flooding, erosion and other impacts of sea level rise.

While all the above structures should be carefully evaluated, it must be made clear that not every option will have widespread applicability in the Wider Caribbean. For it is not the efficacy of the strategies themselves, but rather the various physical, social and economic constraints which dictate their implementation. Some of these factors will be evaluated in the ensuing section.

### 5.2 Limiting Factors To Adaptive Responses In The Wider Caribbean

5.2.1 Retreat: The constraints imposed by physical size would, without a doubt, render retreat an impossible option for much of the insular Caribbean. Table 2 indicates that over 65% of the islands have areas less than 450 sq. km., while nearly 80% do not exceed 1000 sq. km. in size. The table becomes even more revealing, when it is considered that the Bahamas for instance has a total area of 13,942 sq. km., but comprises over 700 tiny islands, only 40 of which are considered 'habitable'. Moreover these islands are rarely more than 120 metres wide and their highest elevation seldom exceeds 60 metres.

Consider also an island like Barbados with a population of approximately 257,000 crammed into an area of 430 sq. km. The population density, 598 persons/sq. km., is one of the highest in the world, and there is already tremendous pressure exerted by the competing land uses for scarce 'useable' land. Clearly large-scale retreat in islands like Barbados, Bahamas and Turks and Caicos could not be considered an entirely viable option. Indeed for such countries, emigration to other nations both within and outside the region, must be anticipated as part of the process of retreat.

TABLE 2
Size of selected countries of insular Caribbean

| Country | Size | Country | Size |
|---|---|---|---|
| Anguilla | 91 | Jamaica | 11424 |
| Antigua & Barbuda | 440 | Martinique | 1080 |
| Aruba | 181 | Montserrat | 102 |
| Bahamas | 13942 | Saba | 13 |
| Barbados | 430 | St. Eustatius | 21 |
| Bonaire | 311 | St. Kitts & Nevis | 269 |
| B.V.I. | 150 | St. Lucia | 616 |
| Cayman Islands | 260 | St. Maarten | 41 |
| Curacao | 544 | St. Vincent & Grenadines | 388 |
| Dominica | 750 | Trinidad & Tobago | 5128 |
| Guadeloupe | 1373 | Turks & Caicos Islands | 417 |
| Grenada | 345 | U.S. Virgin Islands | 342 |

And while it might be possible for the larger states to implement the retreat option within their own jurisdictions, there may be certain environmental and other 'costs' incurred. At present the pattern of settlement largely reflects the influences of topography, the presence of forests and woodland, availability of water and ease of communications. In the process of resettlement, some of the remaining forest would have to be cleared in places like Belize, Costa Rica, the interior of Guyana, Mexico, Honduras, Cuba, Jamaica and Venezuela. In addition the establishment of basic infrastructure such as roads and utilities could be extremely costly, particularly if the terrain is rugged and remote.

Thus, in the context of the Wider Caribbean, retreat would be at best, a difficult option, and in many instances it could be physically impossible to pursue.

5.2.2 **Protection:** Given the central role of coastal and marine resources to the social, economic and cultural life of the Wider Caribbean, serious consideration must be given to protection. Many countries of the region, particularly the small island states, are so dependent on the coast, that retreat would threaten the very foundation of their economic systems. In countries like the Bahamas, Antigua, Barbados, Turks and Caicos and British Virgin Islands where tourism is the prime foreign exchange earner, the economies would face collapse, if tourism plant and associated infrastructure were to be lost.

While protection will be mandatory for many countries, it must be stressed that it could be a very costly option. It is estimated for instance, that the cost of protecting the Caribbean islands from a 1-metre rise in sea level in 100 years is approximately US$ 411.1 billion, at an annual cost of 0.2% of GNP. [38]. The figure covers only the cost of preventing inundation alone and does not include the cost of meeting current coastal protection requirements. The estimate does not include the value of ecosystems that could be lost on account of protection, or the cost of preventing saline intrusion, nor does it cover mitigation of impacts of higher storm frequency. Moreover, the calculation is based on current costs, and assumes that flood risks remain constant[3], it ignores externalities and assumes no change in geomorphological, social or economic conditions [37]. In addition the estimate does not include any annual maintenance costs which would inevitably be incurred.

Notwithstanding the projected high costs, regional governments must be cautioned not to ignore protection as a viable strategy for combatting sea level rise. Indeed there is presently so much at stake for many countries, that protection will be almost mandatory in certain cases.

It is clear then that once protection is chosen as a strategy, a careful analysis of the relative benefits to costs must be vigorously pursued. For given the high structural costs involved, rational decisions will have to be made as to which coastal segments should be protected by the most cost-effective measures. Admittedly the initial costs of protection would impose a financial burden on island economies. But immediate planning for such an eventuality, combined with technical and other assistance from donor countries and international agencies could provide a solution. Perhaps also regional governments may have to consider a reduction in spending on other areas of activity, which yield fewer social, economic and environmental benefits, than those accruing from protection of the coast.

---

[3] The assumption would be that an area which now floods once every 10 years for example, would continue to experience flooding at the same frequency, with a 1 m sea level rise.

5.2.3   Accommodation:   Without a doubt, certain accommodation measures could prove to be efficacious in the Wider Caribbean, which need not incur unmanageable costs.   In the first place high-risk areas need to be clearly identified and mapped, so that strict development controls can be applied.   There is also the need to develop and implement appropriate building setbacks as is currently being done in countries like Barbados, the British Virgin Islands and along the Florida coast.   In many cases this may not require new statutes, but simply amendments to existing ones.

Sea level rise of the magnitude predicted, will also present opportunities both for exploring new areas of activity as well as optimising resources which are not being fully explored.   Aqua-and mariculture are areas of great economic potential for the regional and extra-regional markets.   The region's agriculture also has the capacity to adapt to sea level rise to some degree. Surely our scientists at the various institutions across the Caribbean have the technical competence to develop halophytic and hydrophytic varieties of many crops.   Or alternatively,   suitable   commercial   replacements   for existing crops may have to be considered.

Since the integrity of the region's water resources must be ensured at all costs, various strategies must be expeditiously   researched   with   a   view   to   swift implementation.   Strategies including the relocation of coastal wells and the improvement of water storage facilities (including construction of tanks), should be regarded as priorities, even if predicted sea level rise does not materialize.   The control of drawdown on coastal aquifers   is   another   area   which   should   be   urgently addressed.   Uncontrolled pumping will in fact increase the susceptibility of these aquifers to saline intrusion. Clearly then,   the   implementation   of   effective   water resources management policies, tailored to the specific needs of individual states, is not a negotiable option. And since water resources availability impacts on every sector of society,  vigorous education of the multiple publics about the necessity for water conservation,   must be a vital part of the programme.

## 6.0                       THE WAY FORWARD

While it is anticipated that the Wider Caribbean may be able to cope with some of the effects of the predicted rise in sea level, the only effective insurance against disaster is proper planning for sustainable development. Thus the goal of every nation within the region must be the   implementation   of   a   comprehensive   Coastal   Zone Management Plan (CZMP), which seeks to incorporate the uncertainties of sea-level rise.   Such a plan should take cognizance   of   the   wide   range   of   components   which constitute the coastal system (figure 4).   Scientists now

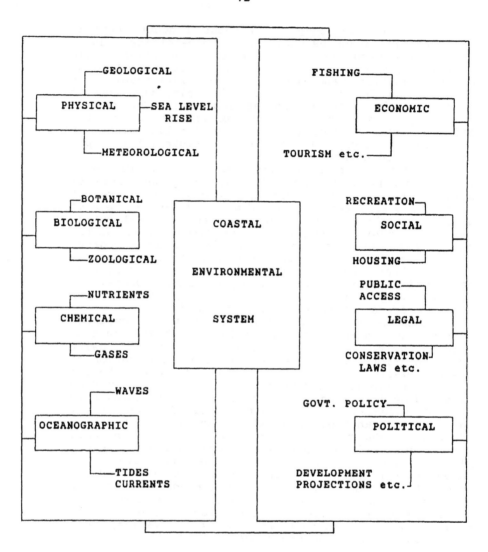

Figure 4.    Schematic diagram showing components of the coastal system which should be taken into account in comprehensive coastal zone management.

agree that many of the likely consequences of a 20 m rise in sea level can be adequately mitigated through sound coastal zone management [16,36,37,39]. Some countries have already begun to embark on this exercise.

Barbados for example began the process in 1983 with the establishment of a Coastal Conservation Unit which undertook a US$1.3 million Diagnostic and Pre-feasibility Study for coastal management. The second phase, the Feasibility Studies and pilot projects implementation, commenced in July 1991 and is scheduled to last over a three-year cycle, at a cost of US$7.3 million. The ultimate goal is the development of a comprehensive Coastal Zone Management Plan. The entire programme is being jointly funded by the Government of Barbados and the Inter-American Development Bank. The Barbados initiative is a clear recognition by Government that the resources of the coastal zone are a mainstay of the national economy. It is widely anticipated, that other countries will follow the Barbados lead, as the threat of a rising sea level now renders sound coastal zone planning more urgent than ever before.

Notwithstanding the extensive literature that exists on sea level rise, there is still much uncertainty about the rate and magnitude of predicted change. Indeed some doubt has been raised as to the applicability of global and hemispheric means to the regional situation [9]. To this end, an appropriate network of tide and water-level gauges needs to be established so that sea-level fluctuations can be more reliably predicted. The current distribution of monitoring stations in the region is not adequate, with coverage in the insular Caribbean especially poor (see table 3). The Inter-Governmental Oceanographic Commission Sub-Commission for the Caribbean and Adjacent Regions (IOCARIBE) has already proposed such a network for the region and it is intended that the data will be fed into the Global Sea-level Observing System (GLOSS). The operation of the system will however call for national entities to conduct the monitoring. They will then be responsible for transmitting the semi-processed data to some central data bank, such as the Permanent Service for Mean Sea-Level (PSMSL). Some countries may not at present have the facilities for undertaking such a programme, so that technical and financial assistance will have to be sought.

Research and data gathering are two of the tasks which should be pursued swiftly. The first order of work must be the development of appropriate vulnerability indices. This would allow those areas and systems (natural and man-made) at greatest risk from sea level rise, to be properly identified and mapped. To develop these indices, site-specific, detailed case studies will have to be undertaken so that appropriate response strategies would be based on local reliable, scientific information. This is not to imply that the region should

ignore the results of global, empirical research.
Climate change and sea-level rise are complex, global
problems; thus the region must continually seek to become
part of the world-wide network which accesses and
exchanges relevant scientific data.    In the final
analysis however, the choice of options and response
strategies must be informed by the results of local
research.

TABLE 3
Tide gauges in operation in wider Caribbean

| Country | No. of gauges | Country | No. of gauges |
|---------|--------------|---------|--------------|
| *Barbados | 1 | Jamaica | 1 |
| Colombia | 2 | Mexico | 7 |
| Costa Rica | 1 | Panama | 1 |
| Cuba | 1 | Puerto Rico | 3 |
| Dominican Repub. | 2 | Trinidad & Tobago | 1 |
| Guatemala | 1 | U.S.A. (Gulf) | 24 |
| Haiti | 1 | U.S.A. (Atlantic) | 6 |
| Honduras | 2 | Venezuela | 7 |

* Deployed on April 9, 1990

Source: Permanent Service for Mean Sea Level, U.K.

There is a growing suspicion among Caribbean
scientists, that the severest effects of sea level rise
will be felt as a result of changes in the
characteristics of extreme events [33,40]. While severe
hurricanes and storm surge, winter swells and tsunamis
are relatively rare, no country in the Wider Caribbean
falls outside their spheres of influence.  These events
cause serious flooding, coastal erosion and significant
alteration to critical marine ecosystems, such as coral
reefs and seagrass beds.  The study of these phenomena
against a background of sea level rise, must therefore
be considered an area of necessary research. This effort
must not only seek to improve current knowledge of the
structure and dynamics of these systems, but should also
aim to develop reliable models for prediction of impacts
on specific ecosystems, in quantitative terms.
Finally, an adequate institutional, legal and policy
framework must evolve to complement and facilitate the
goals of whatever response strategy is employed.  The
success of any option depends not only on technical
soundness or economic and socio-cultural desirability,
but is equally dependent on a supportive administrative
structure.  The choice of institutional arrangements can
only be properly determined at the individual country
level. For the institutional implications of the various
adaptive responses to sea level rise will vary from

country to country. Regional technocrats and policy makers must therefore insist on the allocation of adequate resources, if viable institutional arrangements are to be put in place.

## REFERENCES

1.  United Nations Environment Programme (UNEP), Action Plan for the Caribbean Environment Programme. UNEP Regional Coordinating Unit, 14-20 Port Royal Street, Kingston, Jamaica, 1987.

2.  Nurse, L.A., Global sea level rise: a statement from the island of Barbados. Small States Conference on Sea Level Rise, Republic of Maldives, Nov.14-18 1989.

3.  United Nations Environment Programme - Economic Commission for Latin America and the Caribbean, Marine and coastal area development in the Wider Caribbean area: overview study, UNEP 1979.

4.  Food and Agricultural Organization of United Nations, Yearbook 1988 Vol. 42, Rome, 1989.

5.  World Meteorological Organization - UNEP-Intergovernmental Panel on Climate Change, Overview and conclusions, climate change: a key global issue, UNEP 1990.

6.  World Meteorological Organization - UNEP - Intergovernmental Panel on Change, Scientific assessment of climate change, WMO-UNEP, 1990.

7.  Hoffman, J. S., Estimates of future sea level rise. In Greenhouse effect and sea level rise, ed. M.C. Barth and J. G. Titus, Van Nostrand Reinhold, New York, 1984, pp. 79-103.

8.  Barth, M. C., and Titus, J. G., Greenhouse effect and sea level rise, Van Nostrand Reinhold, New York, 1984.

9.  Maul, G., Implications of climatic changes in the Wider Caribbean region - preliminary conclusions of Task Team of Experts, Caribbean Environment Programme Technical Report No. 3, UNEP-CEP, Kingston, Jamaica, 1989.

10. Delft Hydraulics, Criteria for assessing vulnerability to sea level rise: a global inventory of high risk areas, UNEP 1989.

11. Nurse, L.A., The deterioration of Caribbean coastal zones: a recurring issue in regional development. In Sustainable development in the Caribbean, eds. J. Cox and C. Embree, Institute for Research on Public Policy, Nova Scotia, 1990.

12. Day, J.W., and Templet, P.H., Consequences of sea level rise: implications from the Mississippi Delta, Coastal Management, 1989, vol. 16, 241-257.

13. Vellinga, P., Sea level rise, consequences and policies. In Sea level rise, a selective retrospection, ed. P.C. Schroder, Delft Hydraulics, Netherlands, 1988, pp 1-16.

14. Government of Barbados, Coastal conservation study pre-feasibility report, Bridgetown, Barbados, 1984.

15. Hendry, M.D., Sea level movements and shoreline changes in the Wider Caribbean region. In Implications of climatic changes in the Wider Caribbean region, eds. J. Milliman and G.Maul, Edward Arnold, London (In press).

16. Lewis, J., Vulnerability of small states to sea level rise: sea defence, adjustment and preparedness; requirements for holistic national and international strategies, Small States Conference on Sea Level Rise, Republic of Maldives, Nov. 14-18, 1989.

17. Ojo, O., Sociocultural implications of climate change and sea level rise in the West and Central African Regions. In Changing climate and the coast, ed. J.G. Titus, UNEP/WMO/AM. Soc. Civ. Eng./EPA/NOAA, 1990.

18. Pernetta, J., and Sestini, G., The Maldives and the impact of expected climatic changes, UNEP Regional Sea Reports and Studies No. 104, UNEP, 1989.

19. Miller, D.L.R., and Mackenzie, F.T., Implications of climate change and associated sea level rise for atolls, Proc. 6th Intl. Coral Reef Congr., 1988, Vol 3, 519-522.

20. Ali, M., Sea level rise - a coral atoll perspective on the terrestrial environment, Small States Conference on Sea Level Rise, Republic of Maldives, Nov. 14-18, 1989.

21.  Maniku, M.H., Sea level rise - a coral atoll perspective on the marine environment, Small States Conference on Sea Level Rise, Republic of Maldives, Nov. 14-18, 1989.

22.  Vincente, V.P., Singh, N.C., and Botello, A.V., Ecological implications of potential climate change and sea level rise in the Wider Caribbean, Meeting of Experts on Caribbean Environment Programme, Mexico, 7-9, Sept. 1988.

23.  Neumann, A.C., and Macintyre, I.., Reef response to sea level rise: keep-up, catch-up or give-up, Proc. 6th Intl. Coral Reef Congr., 1988, vol 3, 105-110.

24.  Brown, B., Possible effects of sea level rise on corals and reef growth, Small States Conference on Sea Level Rise, Republic of Maldives, Nov. 14-18, 1989.

25.  Macintyre, I.G., Burke, R.B., and Struckenrath, R., Thickest recorded Holocene reef section, Isla Perez core hole, Alacran Reef, Mexico, Geology, 1977, 5, 749 - 754.

26.  Montaggione, L., Holocene submergence on Reunion Island (Indian Ocean), Ann. S. Afr. Mus., 1976, Vol.71, 69-75.

27.  Buddemeier, R.W., and Smith, S.V., coral growth in an era of rapidly rising sea level: predictions and suggestions for long term research, Coral Reefs, 1988, Vol.7, 51-56.

28.  Wells, S.M., ed. Coral reefs of the world vol. 1, UNEP/IUCN, 1988.

29.  Dubois, R., Coastal fisheries management: lessons learned from the Caribbean. In Coastal resources management : development case studies, eds. J.R. Clark, Renewable Resources Information Series, Coastal Management Publication 3, Research Planning Institute Inc., Columbia, N.Y., 1985.

30.  Glynn, P.W., Widespread coral mortality and the 1982-83 El Nino warming event, Env. Conser., 1984, Vol. 11(2), 133-146.

31.  Woodroffe, C.D., Mangrove swamp stratigraphy and holocene transgression, Grand Cayman Island, West Indies, Marine Geology, 1981, Vol. 41, 271-294.

32. Ellison, J.C., Possible effects of sea level rise on mangrove ecosystems, Small States Conference on Sea Level Rise, Republic of Maldives, Nov. 14-18, 1989.

33. Gray, C.R., Climate change - facts and impacts, National Consultation on Environment, Kingston, Jamaica, 1990.

34. Caribbean Tourism Research Centre, North American demand study for Caribbean tourism, Barbados , 1981

35. De Sylva, D.P., Human health, quoted in Maul, G., 1989.

36. Titus, J.G., Greenhouse effect, sea level rise and coastal zone management, Coastal zone management Journal, 1986, Vol. 14, 147-172.

37. Intergovernmental Panel on Climate Change, Strategies for adaptation to sea level rise, UNEP/WMO, November 1990.

38. Rijkswaterstaat - Delft Hydraulics, Sea-level rise: A world-wide cost estimate of basic coastal defence measures, Netherlands, February 1990.

39. Sestini, G., Jeftic, L., and Milliman, J.D., Implications of expected climate changes in the Mediterranean region: an overview, UNEP Regional Seas Reports and Studies No. 103, 1989.

40. Shapiro, L.J., Impact of climate change on hurricanes, Meeting of Experts on Caribbean Environment Programme, Mexico, 7-9, Sept. 1988.

# SEA LEVEL RISE IN THE MEDITERRANEAN REGION
## LIKELY CONSEQUENCES AND RESPONSE OPTIONS

G.SESTINI

Applied Earth Sciences Consultant
(UNEP Adviser)
c/o Via della Robbia 28, 50132 Firenze, Italy

## SUMMARY

The experience of impacts of periodic high sea-level and exceptional weather events is well recorded in the Mediterranean, especially in relation to the present state of degradation of many coastal and shallow marine areas, involving beach erosion, water pollution, man-induced land subsidence and ecosystem disruption.

Speculation on the effects of an accelerating sea-level rise in consequence of 'greenhouse effect warming', is justified, despite large uncertainties in climate change prediction, because the impacts could be large and far reaching, notably because of increasing occupation of coastal areas due to demographic pressure and economic development.

Substantial climatic change and sea-level elevation would spread and intensify all our current coastal zone environmental problems. The most serious negative consequences for coastal (in some cases, national) economies would derive from impacts on tourism (beaches, marinas); ports, urban seafronts, shore pro-tection structures; roads, railways, airports and industries by the sea, especially those built on lands at or below sea-level; reclaimed-land agriculture and irrigation systems (flood-ing and salinization); coastal water resources; lagoonal fishing.

Although over time, coastal activities may change due to reduced profitability, and to wider-ranging social and economic shifts, the danger (esp. with an accelerating sea-level rise) is that adaptation costs would become so widespread, and escala-ting, that economic stringency and coastal/national conflicts of interests would lead to grave socio-economic upheavals, includ-ing the forced abandonment of some areas/uses and population displacements.

In order not to have to face disasters at short notice and to mitigate conflicts between environment and society, it seems imperative that uses of coastal zones and shallow seas during the next 20 to 30 years (whether at a primary or secondary level development), are rationally planned and managed, on the basis of of sustainable development, intersectoral approach. Much research is needed to identify high risk areas. Environmental impact assessments should henceforth also include probabilistic scenarios of climate warming and sea-level rise.

## INTRODUCTION

At this time towards the end of the 20th century, many Mediterranean coastal areas are experiencing unprecedented change through a combination of natural processes and increasing man-induced impacts. A range of environmental problems, from surface and groundwater pollution and salinization to land subsidence, shoreline erosion, disruption of wetland and shallow marine ecosystems, reduction and deterioration of wildlife habitats [68,78], has arisen in consequence of haphazard, unplanned development and of exploitation carried out in disregard or ignorance of natural systems.

The trend towards expansion of coastal areas use is, however, likely to intensify, owing to the pressure of population growth and of economic development, especially in the countries of the southern and eastern Mediterranean. There is no doubt that in the early decades of next century the weight of increasing anthropic occupation will further damage many portions of Mediterranean littorals. Fortunately, in recent years problems like resources depletion and declining environmental quality have begun to capture widespread interest. There is much greater awareness that appropriate action must be taken.

Sea level in the Mediterranean is areally and temporally quite variable, due to geological factors, and to tidal and meteorological forcing. Storm surges in association with spring tides, generating water levels 1-2 m above normal MSL, have caused and are causing considerable damage to a a number of anthropic coastal activities and land uses, especially in low-lying deltaic areas.

At present consideration of the consequences of sea-level rise in the next century is dependent upon one basic question: will the level rise substantially due to global warming and major climatic changes? There are still many uncertainties regarding the detection, timing, entity and feedbacks of greenhouse warming, and in fact, despite the widespread acceptance of the theory, there are as yet no definite signals of a climate change in act, nor of an acceleration of sea level [56,72,87].

Nevertheless, sea level increase up to and over 100 cm during the next century are a possibility which must not be ignored. The consequences for the Mediterranean coastal zone would be so far reaching, that it is not too early to examine their presumable nature. The region has about 40.000 km of coastline, a quarter of which is around the islands. There is a high concentration of population by the shore, with many important cities, harbours and industrial centers; many cities and archeological sites of inestimable cultural value. Beaches are extensively used for recreation, attracting millions of visitors in summer,from within the region and without, and have become a relevant source of income for several countries, and especially for islands (e.g. in Greece, Spain; Cyprus, Malta).

The Mediterranean margins include as well many lagoons and wet-lands of great economic and ecologic importance. These transitional environments would be most vulnerable to atmospheric warming and sea-level rise, because of their dependence on delicate balances between land and sea.

The main consequences of occasional, or cyclically higher sea-levels at low-lying and particularly exposed coasts are: 1. Storm surge erosion and flooding, overtopping of the more exposed, fixed structures. 2. Higher salinity in lagoons, river estuaries, drainage canals, in groundwater. 3. Derangement and loss of transitional ecosystems. 4. Disruption of infrastructures that are not directly exposed to the sea,

In the time scale of one or two generations, atmospheric warming and accelerated SLR would have implications in different spheres, such as the environment (loss of nature areas, ecological changes), economic (property and production losses, costs of consequential measures), social (unemployment, resettlement, loss of public utilities), and legal-adminstrative (e.g. jurisdiction, and boundary problems). Higher sea-levels would affect many long-term coastal management projects (e.g. water resources, coastal engineering, communications and energy planning, nature conservation).

While exceptional sea levels have highlighted the relation between coastal degradation and vulnerability to natural hazards, the issue of long and short-term climatic change has helped to focus on environmental problems, and stimulated a discussion of coastal zone issues, including planning and management.

A preliminary in-depth assessment of the consequences of climatic change and SLR for environment and society in the Mediterranean region has been carried out by a UNEP-sponsored group [29,68], in parallel with a similar study on the Caribbean region [42]. The cases of the Ebro, Rhone Po, Axios, and Nile deltas have been examined in detail [8,22,40,62]; different aspects of sea-level rise have been discussed by Milliman [46], Flemming [17] and Pirazzoli [55,56,58]. Other reviews have treated the impacts on low-lying alluvial-deltaic coasts [30, 65], on lagoons and wetlands [66], and all the main issues that are related to sea-level rise [67].

Most studies have stressed the role of anthropic impacts on the coastal zone, as being more relevant and certain than those of sea-level change, at least during the next 2-3 decades.

The aim of the present summary is threefold.  To examine the question of sea-level change and of assessment of its consequences; to speculate on the possible impacts of accelerated sea-level rise (ASLR) during the next century on the different types of coastlines and coastal uses in the Mediterranean; to consider what could be sensibly done at this stage, and suggest which activities could be pursued in order to identify localities and economic activities that are potentially at risk.

## GLOBAL WARMING AND SEA LEVEL RISE

There is at present a consensus in the scientific community that a global warming is likely, and on an unprecedented scale, if the build-up in the atmosphere, mainly through man-made emissions, of radiatively active trace gases ($CO_2$, $CH_4$, $N_2O$, CFC's) is allowed to continue [60].

On a "business-as-usual" (BUA) scenario (i.e. unmitigated persi-stence of the present patterns of energy consumption) the assumed con-sequence of a doubling by as early as 2030, is an increase of average global temperature, within a century, of between 1.5° and 4.5° C ('best guess' of 2.5°C); This 'warming commitment' would imply ave-rage global temperature rises of 0.2°-0.5°C each decade, to reach at least 1°C by 2050 [28,81].

Although overall climatic change in the next few decades might be gradual, the atmosphere is expected to become increasingly unstable, with greater frequency of 'extreme' perturbations [85]. Of considerable practical effect  would be a greater recurrence of exceptional weather events, like hot summers and droughts, mild winters with unusual cold snaps; sporadic heavy rains with exceptional river floods; high storm and tidal surges [84,85]. It is these unusual 'crescendo' events [16] that would have the greatest impact on the coast.

In southern Europe one of the main consequence of global warming would be a northward shift of climatic zones, increased evapotranspiration, and changes of air circulation and precipitation patterns. These changes would affect water resources, soils, wetland ecology, coastal stability, and such dependant economic activities as agriculture, fishing and beach recreation.

Sea level could rise (SLR), on account of oceanic water expansion and of the melting of middle latitude mountain glaciers [86]. The most recent,  estimated mean values appear to converge on a figure of about +20cm by 2025 or 2030 and of +65 cm by the late 2000's [69,81]. Nevertheless SLR could range in late 2000's from +30 cm to + 110 cm (low vz high scenarios [28]).

However the actual rise would almost certainly not be the same everywhere world-wide, due to, probably large scale modifications in ocean circulation, a facto quite relevant in the Mediterranean, with its morphological and climatic diversity. Speculations of consequences are still very difficult to elaborate at this stage [20]. How would the denser saline Mediterranean waters expand, in relation to the Atlantic? Possibly warming would alter the relations between the deep-surface-intermediate waters and exchanges at sills and upwellings [90].

It is fair to say, that at this time there are no signals that either climate or sea level have definitely started to change as a result of a 'greenhouse effect' enhancement [81]. Therefore, given even the present impossibility or inadequacy to forecast major storm events or any complex climatic state, Gneral Circulation Models (GCM) and SLR scenarios can only be taken as an indication of the range of possible changes that might occur. Nevertheless, some potential consequences can be identified on the basis of the degree of probability of the different scenarios.

## THE MEDITERRANEAN COASTAL ZONE

Consideration of the physical and economic consequences of a sea-level rise enhancement in the Mediterranean must take into account the diversity of coastal morphology, climate, hydrology and of anthropic occupation of the region. The geologically recent shaping of the circum-Mediterranean lands has produced varied coastlines with mountains by the sea, large to small alluvial plains, fault-bound embayments, islands. Most stretches of high coast are in fact alternations of rocky cliffs, small bays, narrow plains.

Low-lying shores are more extensive than would appear at first [4], Most plains however have a relatively high gradient, zones under 2m elevation being generally narrow. Only in some of the larger deltas (Po, Nile) are there extensive areas between sea level and the 1 m countour line (in the Nile delta, 35 % of the surface lower than 15 m). Reclamation of deltaic marshes and lagoons has in many areas accelerated land sinking. In the NW Adriatic and western Nile delta, parts of the coastal plains now lie up to 4m below sea level, and have to be protected by extensive dyke systems.

The demographic and economic importance of the Mediterranean coastal margin of alluvial plains has increased substantially in the last 100 years (though in some countries more so than in others), after centuries of neglect mainly for safety and health reasons. Present coastal functions and uses include: ports and sea transport, towns (residential/ commercial), recreation/ tourism, industries, agriculture, fishing/ aquaculture, waste disposal, oil/gas extraction.

The importance of the Mediterranean seafront in relation to the rest of the country varies; for instance it is relatively less so in Spain, France and Turkey than in Italy, Greece, Albania, Algeria, Israel. In Greece [41] as much as 90% of the population lives within 50 km of the coast and all major industrial centers are coast-related as well as much of agriculture. In Egypt, the Nile delta north of Cairo represents 2.3% of the area of the country, but contains 46% of its total cultivated surface and 50% of its population; the belt 0-3 m harbours about 20 % of the population (with Alexandria 3.5 mill., Port Said 450.000 inhabitants), 40% of industry, 80% of port facilities, 60% of fish production [64].

The Mediterranean coastal lagoons and wetlands (e.g. Laguedoc-Rhone delta, NW Adriatic, Albania coast, north Greece deltas, Ceyhan-Seyhan delta, Nile delta, north Tunisia) are special environments [66] because of a primary productivity that is much higher than in the sea, and of ecosystems that rely on a delicate balance between brackish and freshwater conditions. Lagoons and wetlands (esp. tidal ones) are of great biologic significance for the fauna of adjacent estuaries and marine areas and are essential habitats for migrating and nesting birds. They are an important resource, as nursery grounds for many commercial fish species; as habitats for wildlife and as zones of nature conservation. Lagoons and marshes also form a natural pollution control mechanism to cleanse inland waters.

Historically, fishing in the Mediterranean has always been an important social and economic activity, quantitatively prevalent in the coastal zone. It has declined in relation to other occupations, except in some parts (e.g. Tunisia, Greece, Spain, Italy); overall production now falls 2/3 short of demand. The potential for aquaculture is considerable, there is a long tradition of breeding fish and shellfish, a favorable climate and environments [68].

Coastal tourism is currently the greatest consumer of the Mediterranean littoral, exploiting the edges of coastal plains as well as the cliffed shores. Utilization is high in Italy, Spain, Greece, France (to over 75% of total coastline). It is rapidly expanding in Egypt, Turkey, but it is still minor in Algeria, Libya, and Albania. Not all tourism is beach-oriented, in some countries 'cultural' tourism is equally, if not more important (e.g. Italy, France, Egypt, Israel).

The relevant contribution to GNP's, (foreign exchange, employment has encouraged ambitious tourist development plans in some countries. Pursue of the beach recreation market has involved the construction of hotels, apartment blocks, holiday villages, second homes, caravan parks, camping grounds, pleasure boat harbours and related infrastructures and services (roads, parkings, shopping centers). Some resorts are very urbanized and old towns have expanded, often with a loss of identity. Coastal tourism is concentrated in a narrow space and time (two summer months), thus it places high stress on water resources, waste disposal systems, on the environment and landscape; it is culturally, and often also economically detached from the hinterlands [15].

In all Mediterranean countries there is a high density of population near the coast, whether by the shore or within 10-50 km [77]. In several countries (e.g. Algeria, Tunisia, Greece, Libya, Israel, Italy) over 50% lives within 50 km of coast (100% in Cyprus, Malta, Lebanon). Consequently, the urbanization of coastal areas is intense: 45-50% of the population. Lack of space in some cases has had to be managed by expansion of land by filling (e.g. Monaco, Genoa) and by building roads and railways quite close to the sea (e.g. Riviera, Côte d'Azur, SW Spain).

There are at least 360 cities and towns over 25.000 inhabitants that are situated by the sea. With a few notable exceptions (e.g. Venice) only small portions of most major cities lie between sea level and one meter above, in any case protected by 2-4m high seawalls (Barcelona, Marseille, Nice, Naples, Algers, Tripoli, Alexandria [12]). Apart of cities, the actual coastal fringe 0-1/2m has a low density of resident population [1]. In summer however, the population of some beach resorts increases 10-15 times [7,63].

A legacy of the past reliance on maritime communications is the large number of ports, natural and man-made, large and small. Nowadays they explete a number of different roles, from the traditional ones of national export-import and local transportation (e.g. Aegean, Tyrrhenian seas), to international transshipments (e.g. Genoa, Trieste, Salonike, Volos, etc.), to passenger and automobile ferries, pleasure boating, and fishing. A special function is that of oil loading/discharging terminals.

## SEA-LEVEL VARIATIONS

### Factors of sea-level variations

Sea level in the Mediterranean is far from an even surface, which would respond to steric effects, by equal amounts everywhere.  Differences arise [55] not only from general factors of world ocean surface inequalities (eustatic, glacio-hydro-isostatic, climatic and tectonic  but also from characteristics that are proper to the Mediterranean: the narrow Gibraltar connection with the Atlantic, an unbalanced water budget, earth density effects and recent geological activity, distinctive water circulation features (gyres, jets, meandering currents) which reflect a complex, three-dimensional geometry of basins and sills [35].

Sea-level changes randomly or periodically (with intra-diurnal, diurnal annual, decadal cycles) in response [32] to: variations of sea-surface temperatures, river discharge, atmospheric pressure, winds and the currents they generate, rainfall, surface water density (e.g. in connection with cold-water upwellings), and tidal forces. Expansion and contraction of shallow water bodies can produce 5-10 cm  sea-level changes within weeks of air warming or cooling [86].

In the Mediterranean, sea-level tends to be relatively higher, seasonally, in the North than in the South [46] (in general river run-off may account for 20-40% of sea-surface changes [32]). A significant portion of the annual change in sea-level can be ascribed to  variations in atmospheric pressure, in the order of 5-15 cm [35].

Quantitavely more significant sea-level shifts are induced by winds, wind-related storm surges and (at some places) spring tides. In the Gulf of Lion [8] sea-level can rise by 1.8m with SE winds, or be depressed to -0.50 m  by the onshore NW winds. Storm surges in the Mediterranean raise the level of 1 to 5m winter waves by 1.5-2.5 m [62,65]. Mazzarella and Palumbo [44] found that the sea-level maxima are more significant and several times larger than MSL variations; their analysis provides a more realistic and adequate basis for compi-ling probabilities of extreme sea-level events in coming years.

The tidal excursion is generally low [46], with spring tides of 20-30 cm (e.g. Ebro, Rhone deltas, North Tunisia, Aegean and Eastern Mediterranean). They highest tides are in the Gulf of Gabes (up to 2 m), in the NW Adriatic (60-90 cm), and the narrow bays of the Turkish coast (Iskenderun, Izmir: 50 cm). Exceptio-nal tides, however, rise to at least twice these amounts. At the Nile delta coast spring tides can reach 80-95 cm (every 20 years), even 120-130 cm (every 50 years) [69]. In the SE Mediterranean high tides occur in combination with storm surges  at least twice a year; sea-level can then be significantly higher, given wave setups of 1.5m  over winter storm waves of 1.5-3 m heigth [51].

A more gradual and longer term factor of sea-level change are geological movements [13,17,55]. As the couple sea-land is not a fixed entity, but ever-changing in relation one to the other, due to tectonic and sedimentary factors, sea level is a relative concept (RSLR).

Subsidence at deltas, whether natural or man-induced, can significantly contribute to sea level rise [48]. In the Po delta, Italy and in the eastern part of the Nile delta, Egypt, the estimated rate of subsidence is about 5 mm/year [64,71]. In Venice, it is presently of 1.33mm/y [56]. In Alexandria (Egypt) 20% of 2 mm/y RSLR (1958-88) is due to subsidence [12], at Port Said 3.6 mm  out of 4.8 mm/y [13].

Locally, land subsidence has been further accentuated by water (and probably by natural gas and oil) extraction [6,48]. In the Po delta excessive groundwater pumping caused during the 60's 70's, a subsidence of up to 350cm. After pumping was prohibited, local sub-sidence has gradually returned to its natural rate. However the ground elevation loss has remained [63]. In Venice between 1930 and 1970 RSLR reached 24 cm, 4 cm being due to natural subsidence, 9 to eustatic SLR, the rest to aquifer dewatering [6].

Other parts of the Mediterranean on the contrary, bear evidence of emergence [3,13,14,55] from archeological remains and various raised morphological features (e.g. in Crete, Rhodes, south Anatolia to north Syrian coast, Peloponnesus, Calabria, parts of Algerian and French coasts) tectonic uplift generally has been slow, but more rapid when associated with earthquakes [57]. Commonly  vertical earth movements in the Mediterranean are said to occur at a rate of 1-5mm a year, averaged over centuries, of 3-20mm a year over periods of 15-20 years [17].

## Historical and recent sea-level changes

Curves for eustatic change during last 10.000 years (worldwide) show that the rate of SLR has varied considerably, with periods of acceleration, still-stands and even reversals. Fluctuation during the past 6000 years, and particularly since 2500 BP have been considerable (no less than 44% of published curves indicate emergence) [59].

The rather frequent coastal archaeological remains from Roman times in the west-central Mediterranean generally reveal a submergence of 0.5m since 2000 BP (i.e. 0.25 mm/y SLR), except where accentuated locally (to 1.5 m) in areas of subsidence, or transformed into emergence in those of uplift [17].

It is generally stated that during the last 100 years there has been a global average SLR of about 15 cm [24,47,81]. According to Pirazzoli [58], however, much of the measured rise in the last century reflects local tectonic and isostatic effects at tide gauges that are concentrated in NW Europe, eastern North America and Japan. Along the coasts of Europe, when land movements are subtracted from the tide gauge data, regional MSL changes show an average rise of about 4 cm (0.66 mm/y) from 1890 to 1950, and a pattern of quasi-stability from 1950 to 1980 [58].

In the Mediterranean and Black Sea a number of SL curves from tidal gauge data [43,59] suggest a tendency to SLR, but at variable rates (0.3-5mm/y), and with considerable oscillations within intervals of years or decades, of the order of 5-8cm [35]. There seems to be, also, a general reversal of the rising trend since the early 70's. The shorter-term fluctuations have been correlated with the cyclicity of solar activity [50]. The tide gauge stations are much influenced by relative land movements at their localities (e.g. subsidence at Venice, Sfax, Antalya versu uplift at Izmir, Oran, Marseille, Port Vendes).

## The future of Mediterranean sea-level

As regards the next 10-20 years, extrapolation of the tidal gauge data for the last several decades suggests that there is a time lag of 18-19 years before glacier retreat and sea-water expansion begins to be noticeable after a period of rising temperature [43,58]. After the warming of the 1980's, sea-level might start rising again, perhaps by 1-2 cm over a couple decades; whether it would continue to rise, it will depend on the evolution of global surface air temperature. For the moment it is yet impossible to detect any definite signs of a SLR acceleration that is unequivocal above the noise of interannual variability and local factors.

Sea-level history in the Mediterranean appears to have varied greatly from place to place, in the recent and more distant past. Some predictions of the direction of uplift or subsidence may be possible, however, since at places the (short-time series) tidal gauge data seem to correlate with the data longer-term trends of archeological data [18]. However, because of the many factors involved in sea level variations, no site-specific prediction can be made as to how seawater would actually rise in the future.

In consideration of the current IPCC estimates (and their uncer-tainties) the outlook for the next 100 years could be conceived as follows:

1. Climate will change, but 'as usual', that is, within the variability of the last 2000 years, and particularly of the last 100 years [33]. Sea level will continue to rise, generally at more or less the present rate (1.2-1.5mm/y), but with conside-rable relative differences between parts of the Mediterranean, due to subsidence/uplift, to seasonal or exceptional meteorological factors.

Within 100 years the result could be a relative SLR of: 0 to 15-20 cm at stable or uplifted coasts; of 35-70cm by subsiding plains and deltas.

2. There could be a measure of global warming, but (for a BUA scenario) not on the forecasted scale, owing to various side effects and feed-backs. A SLR rate perhaps double the present one, would lead to augmen-tation of: 0 to +30cm at uplifted or stable coasts, of +65 to +100 cm along subsiding coasts.

3. Global warming and SLR will occur as currently predicted, the entity of increases depending on deliberate controls on radiative trace gases emissions. The amount of relative SLR in the Mediterranean will again vary according to local conditions, at places being small or none, at others twice the expected amount. By 2030 sea-level could rise as much as during the last 100 years; for a middle scenario by the later 2000's stable coasts could experience up to 70cm SLR, deltas up to 120 cm.

## SEA LEVEL *versus* AIR TEMPERATURE CHANGES

The problem of a considerable augmentation of sea level should not be considered independently of its primary cause, that is of atmospheric warming. It is in association with storm and tidal surges that the impact of SLR would principally manifest itself; since storms are related to winds and waves, their entity, frequency and direction would be affected by the possible changes of atmospheric circulation.

In shallow marine areas and in the transitional marine-terrestrial environments, the direct and indirect impacts of SLR would be added to those of air and water warming: increased water salinity, decreased oxygen content, density, temperature, altered fresh water discharges, due to changed inland precipi-tation. Increased evapotranspiration would have a significant impact in North Africa where fresh water inputs are scarce or absent.

In lagoons periodical excessive temperatures, high salinity and less dissolved oxygen, would ultimately cause considerable changes of fauna and vegetation communities, and the disappearance of bird habitats. For example, the function of low-salinity plant areas, as main nursery and feeding grounds of fishes and waterfowl, would be particularly endangered [27].

A likely northward shift of climatic zones would imply longer summer conditions at the expense of other seasons (with all its ecological consequences); it would affect storm tracks, and thus probably precipitation, not only in coastal areas, but also in the related river-drainage basins. Regrettably, scenarios for storms and rainfall are still uncertain, since there is as yet no climate model specific to the Mediterranean, and constrained by actual geography, which could help to estimate the impact of warming patterns on pressure distribution and atmospheric circulation; that is, the future of cyclogenesis [84]. A large scale general warming could reduce the cyclogenesis-promoting, winter temperature contrasts between land and sea in the Mediterranean, and thus could reduce precipitation and storminess. On the other hand, increased evaporation could lead to greater atmospheric moisture, and rainfall by convection might increase instead.

Changed wind circulation, might alter the relative importance of the different directions and fetches of waves, and thus energy at the coasts. Storm waves impact could equally increase, or lessen; however, an intensification is to be expected of the frequency of exceptional high water occurrences from tidal and storm surges [44,70,85].

## THE PROBLEM OF ASSESSING FUTURE CONSEQUENCES

The present uncertainty surrounding the forecast of sea level and oceanic temperature changes makes a fair assessment of the magnitude of the environmental and socio-economic consequences highly speculative. Uncertainty surrounds also the responses of physical-biological systems, given their complex, synergistic nature, and the future level of coastal zone development and occupation, of changes of local/ world economic relations, of social habits (e.g. leisure, transport), and those generated by technological innovations.

Problems with assessment derive, in the first place, from the inadequacies of GCM predictions. Even if GCM scenarios can be integrated with data from past records of warming periods [38], climate systems are too complex to predict, and there are yet large uncertainties, especially at regional level. A major deficiency is the GCM's inability to reproduce present-day regional weather patterns, and to offer consistent predictions of rainfall distribution. Given even the present impossibility to forecast major storm events or any complex climatic state, the current GCM and SLR scenarios can only be taken as an indication of the span of possible changes (of temperature, rain-

fall, SLR) that might occur over broad regions. They are too
vague to be used for area/site-specific assessments.

Secondly, there is the question of the time scale at which
to evaluate the effects of SLR: within 20, 40, 50 years? If
there is an acceleration of SLR, when would the serious problems
materialize? Warming and RSLR rates would become appreciable
(i.e. +1°C and +20-50cm), within 35-40 years; that is, in the
lifespan of next generation. However, during the last 30 years
alone, coastal zone occupation has been so intensive that many
stretches have been completely transformed. In two more decades
the Mediterranean coastal zones undoubtedly will be further
modified by the spread of tourist resorts and defence structures,
by alterations of river water and sediment discharges, by the
side effects of large construction projects with the life of
decades, and by additional urban and industrial pollution.

Thirdly, there is the uncertainty of socio-economic projections. Many studies have used the present level of coastal development as a basis to estimate consequences at the time of 1m
SLR [ref.75], that is, a 100 year's projection! In fact, if the
physical impacts of SLR on coastlines could probably be predicted, even modelled quantitatively, it is far more difficult to
quantify the consequences of the physical and biological changes
on the future state of land uses [65]. In addition rapid population growth and economic expansion of the Mediterranean coasts
will certainly lead to to the further degradation of the environment, and to resources use conflicts, with atten-dant
political and economic tensions [2].

According to the Blue Plan estimates [77] the population of
the Mediterranean countries by the year 2025 could be increased
two-threefold to 430 or 500 million inhabitants, 40-50% of which
in the 50 km coastal band. Urbanization could almost double, to
70-80% (in the south, maybe accelerated by the further spread of
desertification). Urban water usage could increase by 40-50% in
the North, up to 4 times in the South. Overall electricity consumption probably will go up from 110 Twh in 1985 to 900-1000 Twh
in 2025. The number of summer tourists crowding beaches could be
as much as 380 to 760 millions in 2025 (depending on future
levels of relative development), as compared to the 100 millions
of today (unless saturation will eventually produce strong
negative feedbacks).

In terms of economy, a difficulty with evaluating future
situations, is that 'calculating values that result from a given
change, at a given time horizon, ignores any additional changes
in subsequent years' [1]. Such changes would be the economic
value of future development, in function of the policies followed
(e.g. land use control *versus* development without any coastal
management; the cost of doing nothing vz that of anticipating
SLR, vz that of relocation [23]; the impact on property values,
investments, on one hand of an increasing demand for waterfront
land (also due to post-1992 EEC open market policies), on the
other by society's possible negative responses to knowledge that
sea-level will indeed rise faster.

In addition, coastal zone economy of many parts of the Mediterranean is becoming less and less conditioned by local needs, more and more by hinterland and external economic-financial relations and market forces. The future relevance of local industries, agriculture, ports, will largely be conditioned by worldwide commodity prices and trade trends, such as those of mineral and energy raw materials (with their effects on heavy and chemical industries); those of cereals and industrial crops; and by the demand for consumer goods in a competitive interna-tional society. The role of individual ports may change, in response to altered trends of maritime trade (e.g. the Suez canal after an eventual decline of petroleum transport),and their relative importance depending on the degree of adaptation to technological changes.

It is not easy to quantify, now, let alone in the future, the relative value of coastal economies in function of the rest of a, or several, countries. However, 'disruptions to one link in society functional network would spread widely to other links' [23].

Atmospheric warming itself would affect, directly or indirectly some coastal activities that are related to a wider economic context; for instance, derangement of Mediterranean export agriculture to northern Europe; a longer beach tourist season could be offset by generally warmer climates (esp. if the incidence of environmental problems like algal blooms, pollution would become a more frequent drawback).

## AN ESTIMATION OF FUTURE SLR IMPACTS

### Direct impacts on open coasts and lagoons

The direct effects of relative SLR are those of physical impingement of higher water levels and storm surges, and of flooding. Low sandy coasts, that are shaped by marine processes have the capacity to reform themselves after major storms and to rise in phase with the average rise of sea level, gradually migrating inland. Calculations made in the Netherlands have suggested a retreat of 40-75m for a SLR of 0.5m; of 120-225m for one of 1.5m [26]; probably even greater [79], e.g. from 100 to 500m for 1m of SL rise, in estuaries, embayments and barrier island situations.

The retreat of the beach-dune zone, without backshore flooding, essentially depends on the persistence of an adequate supply of sand from rivers, cliff erosion and littoral drift. If the sand budgets are negative, longshore transport insufficient, the dunes are obliterated, and sand removed seawards, coastal erosion is accelerated and the breaching of beach-dune ridges by storm waves and the breakup of barrier islands would become more frequent. Natural or man-made subsidence, and interference with littoral sand movement [5,21,91], would exacerbate the process.

Thus, since sand deficiencies will persist well into the next century [9], the stretches of Mediterranean coast that are unstable or retreating at present [3,8,19,41,52,63,64], would be even more so in 40 years time, whether SLR amounts to 5-6 cm of

20 cm, and in spite (and often because) of fixed defence structures); in particular, all the promontories of subsident deltas in prominent situations (Ebro, Rhone, Po, Arno, Nile), where the rate of land loss would double or treble [10].

Equally, sea-level rise would worsen the often already precarious situation of the beaches of small bays and narrow coastal plains (esp. in islands), where sand moves in a closed system with a scarce supply (ref. Alexandria, Egypt [10]). A very large number of small fishing and pleasure harbours, of small towns, villages by the sea in such locations, would eventually be impaired. Many archeological sites presently at about MSL would be submerged.

As regards coastal lagoons and lakes, under natural condi- tions they could shift landwards if the rate of sea level rise is low [61,74]. Since retreat would be impeded by existing dykes, even a small amount of RSLR (e.g. 10-20 cm) would have a consi- derable impact, because of the wetlands shallowness, with the flooding of the marshes and reed beds that are paramount to aquatic and avian faunas [27].

The consequences of relative SLR have already been borne in several parts of the NE Italian coast (e.g. Ravenna, Venice, Po delta), where regular basements flooding, deterioration of build- ing materials, disruption of underground public utilities has occurred. The Venice lagoon has been affected by higher and more frequent tidal flooding, due the combination of land subsidence, the excavation of deep navigation canals, the reduction of marginal marshes and tidal flats (following land and industrial reclamation); and of the fixation of aquaculture basins margins [21,56].

As Venice lies 0.8 to 1.5m above MSL, a tidal rise of +1.4m floods 90% of the city. The greatest threat to the city is from an increase of storm surge frequency. With a 20cm SLR, the most dangerous storm surge levels would become thrice more frequent; exceptional high tidal flooding (like the 1.94m tide of 1966) could recur every 15 years instead of 165 years. With a 20 cm relative SLR, Saint Mark Squarewould be flooded by 55% of high tides, instead  of 15% as today, by 75% with a SLR of 30 cm [56].

The main threat of flooding and shore erosion would there- fore derive from, even small increases of the recurrence of extreme storm events. Mazzarella and Palumbo [44] calculated that a 20 cm rise would halve the repeat time of the Nov.1969 exceptional surge at Trieste, from 80 to 40 years.

**Direct impacts on fixed structures and artificial shorelines**

In the case of rocky cliffs, sea walls and piers, water would become gradually deeper, with increase wave elevation and impact. Consequences for maritime structures [26] would be overall increase of overtopping frequency for harbour and quay grounds, greater horizontal and vertical loads, greater energy expenditure, lower clearance under bridges, salt intrusion. Also, the stability of parallel bulkhead and seawalls is dependent on protection from a beach sand ribbon, which on a negative sand budget, will tend instead to disappear [5,45].

In the Mediterranean, a general (relative) rise of SL of 20-30 cm probably would affect a large number of fixed structures (breakwaters, sea-walls, dykes, piers), also because low tidal ranges have allowed low clearance of many harbour quays [80].

However, the vulnerability of Mediterranean ports, oil terminals, urban seafronts, roads, railways, power stations by the sea, varies with the degree of relative exposure to storms surges. For instance, there are many natural, very sheltered ports, where wave energy is low, such as Rijeka, Split, Zadar, Pula, Cartagena, Toulon, Brindisi, Izmir, Iskenderun, Lattakia, Larnaca, Rhodos, Valletta, Siracusa, Palma). Other cities and ports have older parts that are quite sheltered, but new extensions that are exposed (e.g. Genoa, Marseille, Algers).

Very exposed harbours are those of Malaga, Alicante, Tarragona, Haifa, Tripoli, Benghazi, Alexandria, Bizerte, Barcelona, Genoa, Ancona, Leghorn, Civitavecchia. A particular situation is that of canal-estuary ports like Valencia, Ravenna, Tunis, Port Said, and of cities (airports, industrial plants) within or by lagoons, which would suffer from flooding, rather than from direct wave attack. Many Mediterranean cities are situated by the sea, but in fact, generally only in small part they lie, each, between sea level and 1-2m; as it is the case, for example of Beirut, Alexandria, Marseille, Naples.

The degree of vulnerability of the Mediterranean coastlines varies also according to coastal orientation *versus* that of waves and their fetches. Storms with high waves (2-6m) are generated in the Western Mediterranean (and to a lesser extent in the central basin) by westerly winds, and by depressions that migrate eastwards as anticlockwise cyclones.

They cause wave paths to E, NE, N, NW in the west Mediterranean, in the Adriatic and Aegean Sea, and towards the SE in the Eastern Mediterranean. The northerly cold and violent winds that blow in winter from central-east Europe (mistral, bora, etesian, poiraz) on the other hand, generate waves that attack the Algero-Tunisian, Sardinian, West Adriatic, central Aegean and Nile delta coasts [8,41,52,63,64]. All other coasts and cities are more sheltered, (especially those of the eastern Adriatic, W and S Turkey, eastern mainland Greece, Syria), and suffer from a smaller degree of wave impact.

### The indirect impacts of sea level rise

The main indirect consequences would be salt intrusion (soils, surface and ground-water), hydraulic disruption, alteration of ecosystems. Salt fronts would migrate inland, in both aquifers and estuaries [ref. 73,76], especially during droughts (with a threat to municipal and industrial water intakes). Unconfined aquifers beneath sandy shores would be particularly vulnerable to salt water intrusion [68,75], e.g. a 1.5m SLR would cause a 30% increase of salt water seepage [26]. As a consequence more freshwater would be needed to flush low-lying agricultural land.

Problems would arise for water drainage management. In the Netherlands adaptations are considered necessary, but natural drainage would still be possible in parts, if SLR rise is not over 0.5 m; more numerous and larger pumping stations would be required [25,26]. This situation would obtain in the NW Adriatic coast and in parts of the Nile delta. The maintenance of the adjacent agricultural lowlands would require the elevation of all marginal dykes and an increasingly polder-type management.

Higher saltwater levels in lagoons, canals, estuaries would certainly accelerate the physical decay of some riverain towns, even if SLR is not more than 10 cm. The remedy would be to regulate the access of sea-water by sluice systems. However, if the lagoon outlet gates were to be kept closed for long periods of time (and esp. during storms) inflowing rivers/ canals would also increase water levels; pollution would increase, salinities would change, with negative effects for fauna and vegetation [10,56].

SLR would also affect all other infrastructures that are not dir-ectly exposed to the sea, but still related to sea-level (sluices-locks, pumping stations, bridges). In addition, coastal plain lowlands by the sea would face greater river-flooding risks in conjunction with storm surges (like it happened in NE Italy in 1964 [63]).

To conclude with, if an average SLR of 10-20 cm *per se* may not be a great threat to both open and artificial shores, and could be coped with, technically, except in subsident areas, subzero lands, lagoons and canals; there is no doubt that the consequences of elevations above 30-50 cm would be serious: many beaches would disappear, possibly some lagoons could be transformed into bays, reclaimed lands would be flooded (direc-tly or indirectly); contemporary sea-fronts of tourist resorts and towns, industrial installations built on lowlands and by lagoonal areas, pleasure harbours and marinas; desalinization facilities (e.g. Malta), all would suffer extensive damange and would require major restructuring.


## THE ECONOMIC AND SOCIAL CONSEQUENCES OF SLR

As the state of coastal occupation and use around the Mediterra-nean is so varied, it is difficult to be site-specific as to the socio-economic implications of SLR. Most affected would be shoreline-specific uses like settlements, harbours, beach tourism (as a social habit, investment industry, national/local income and revenue); cultural-heritage assets like historical towns, archeological remains; nature protection and conservation.

Given the presence of so many fixed assets that would need to be maintained [62], the obvious consequence of higher sea-level would be the cost of protection and adjustment. Many fixed structures (incl. dykes around sub-zero lands) would need

to be raised, or periodically restructured, according to higher
storm surge risk levels; beach nourishment schemes would need to
be intensified; and, now or later, also the blocking of canals/
estuaries by gates.

Estimates made in the Netherlands [26] have envisaged an
expenditure up to $1.5 bill. for adaptation to ensure safety, in
case of a 20 cm SLR; of $ bill. for a 60 cm SLR; of $ 8-9 billions
if SLR were to be 1 m (and as much as $15 bill. if, in addition,
ocean conditions were to become 10% worse than today). A large
proportion of these sums would be needed to adjust dyke systems
[25].

Beach maintainance by dumping sand dredged offshore would
also be quite expensive; in the eastern U.S. it would budget for
$10-150 billion for 5000km of coastline ($1/2 mill./km). Beach
nourishment is a costly operation requiring a continuous investment, because benefits are temporary, as beaches remain stable
only 5-20 years [54]. A further problem in many Mediterranean
areas would be the source of the sand (it is estimated that 1 m
SLR could require 1000-10.000 $m^3$ sand to face with [76]).

These estimates are generalization, often based on arbitrary
assumptions, especially as regards the extent of adaptation and
protection. Nevertheless, they give a measure of economic
impacts. In the coastal Mediterranean countries the cost of
dealing with SLR effects (adaptations at low coastlines,
harbours, beach nourishment, urban seafronts) would vary [49]
from $17 bill. (Italy) and $ 5 bill. (Greece), to $3 billions
(Egypt), 2.9 bill. (Yugoslavia), $2.2 bill. (Tunisia), $1 bill.
(Algeria, Israel) $ 550 mill (Malta), $210 mill. (Syria), $190
mill. (Cyprus). The cost of beach maintenance would be 50% or
more of the total in Greece, Italy, Spain, Yugoslavia; a notably high percentage would have to be spent for harbours and
urban seafronts adjustments in Algeria, Cyprus, Lebanon, Egypt,
Italy, Spain, Israel, Turkey.

The economic cost of adaptation (to a SLR of +20cm by 2030)
would be proportionally less for coastlines and specific sites
that are in more sheltered locations. However, considerable
expenditures would be imposed even by a low RSLR (15-25cm) for
lagoonal management (marginal dykes, hydrology, aquaculture,
e.g. NW Adriatic, Nile, Camargue, Albania, W Greece), lowland
agriculture (for flood control structures, changes to crops,
adapting to salinization, provision of adequate quality water),
for adjustment of water intakes and wastewater systems, for
provision of urban water (from distant sources? [1]).

Relatively to the whole basin, open sea and lagoonal fisheries would not be affected by SLR and water warming, except on
a qualitative way (composition of catches). Rather, the main
impacts would be from anthropic stresses, such as overfishing,
lagoonal management projects, etc. [ref. 10]. The loss of
agricultural surface would be marginal (few tens to hundreds
meters in general), even in deltas (few 100's m to 1-2 km in
particularly vulnerable areas). The main damage to coastal
agriculture would be from soil and groundwater salinization.

Thus far, the economic consequences are considered in the context of the present economic setting, and only of coastal zones. Futher economic impact would be derived from loss of activities: jobs, services, (ref. dependance of maritime cities on their harbours, the islands increasing socio-economic dependence on beach tourism), the cost of altered economic patterns (loss of property values, activities become economic); of implementing zoning in already fixed-up areas; of the relocation of infrastructures/population, if land actually vanishes. The viability of many beach resorts would probably become questionable, as the cost of beach protection and maintenance escalates; thus, an alteration of state revenues from tourism is to be expected.

In terms of national economy and resources, the guesstimated expenses of managing SLR could amount, in all Mediterranean countries, to less than 1% of GNP [49], but the financial impact would be larger (based on present conditions) in Albania, Malta, Tunisia, Lebanon, Egypt, Greece, Cyprus; relatively much lower in Algeria, Syria, Spain, France, Yugoslavia.

Projection of costs may change however, if one considers that more coastal occupation due to population pressure will take place in Syria, Turkey, Israel, Egypt, Libya, and the Maghreb countries (in the form of new towns, land reclamations, power stations, port expansions and oil terminals, power stations) [2].

Coping with SLR consequences would probably be manageable if the rate of elevation is like today's or twice as much (i.e. 1.5-3cm per decade), since expenditures would generally be covered by the normal 10-20 years periodic maintenance and adjustments. It would become increasingly difficult, if the rate of elevation is greater. With coastal saturation in 20-30 years time, irrespective of the level of risk, economic disruption would become more and more diffuse, financial outlays would no longer be site-specific, but would practically spread to all inhabited/used shores.

An important question would then be that of the responsibility for meeting the costs. At present, in most countries the ownership and responsibility for coasts is usually the state's, which has to pay for protection and for the damages caused by natural hazards. Economic necessity may force changes in the authorities attitudes towards legal /insurance cover of local bodies/individuals. Full government funding for shore protection, or insurance, compensation against 'natural disasters', may be replaced by incentives to implement zoning codes and building regulations. In the case of private investments, it would have to be accepted that protection/adaptation expenses are borne by owners, as part of their investment risks.

## RESPONSES AND POLICY OPTIONS

It is common opinion [36,88,89] that the choice of response strategies would depend primarily on economic and social factors, with three alternatives:

1. Protection: to try to hold back the sea (active defense, prevention).

2. Accomodation: to adapt to SLR by means of coastal planning, in order to minimize losses or the cost of adjustment and adaptation.

3. Retreat: to give up the doomed coastal stretches and land uses, or at best, carry out planned abandonment.

Clearly the situation in the Mediterranean countries is rather different than that of the Netherlands and of the eastern United States coasts, where the issue of response-policies has been principally addressed so far [25,73,75,88]. Specific low-lying coasts may be similar, but their relative importance for the rest of each country is much more limited.

What would be a sound approach at present, should anything be done today, if it might turn out to be unnecessary? adopt policies for something that may become significant only in two generations time? On the other hand, if it is assumed that nothing will happen and no action is taken, what would be the consequences if it actually occurs?

Given the current degree of ASLR probabilities, the alternatives may be as follows.

1. To argue that the formulation of preventive/adaptive studies, and adoption of reactive policies are necessary, even 'urgent', because the longer such measures are postponed, the greater their eventual cost [75]. Taking actions today may be justified by the likely savings from preparing for SLR *versus* its very large likely negative impacts [23].

2. To take the view, instead, that policy cannot be based on estimates that have a high degree of uncertainty [37]. An appropriate understanding of the alternatives (evaluation of costs of defense, of altering development patterns) can significantly prepare for eventual action to reduce adverse impacts, if need be.

Extensive analysis is required and debate should precede decisions about which actions to undertake; formulation of response strategies [49] must be based on appraisal of the level of vulnerability in the framework of costs (with the alternatives of no preparation *versus* active preparation for each site, region), and of the degree of probability of each SLR scenario (e.g. no ASLR, +30cm, +50-100 cm).

## Protection

Where investment values are high (either public: cities, ports, infrastructures; or private/mixed: beach resorts, marinas) holding back the sea would probably be cost effective, even if involving considerable adjustments to existing hard protection, and 'soft' protection measures (sand filling) in the case of beaches.

In the Mediterranean region, intensity of coastal occupation
and particularly, the large number of unmovable features (cities,
ports, valuable agriculture) suggest that primary reliance to be
placed on raising bulkheads, re-designing seawalls [ref. 36].
Responses would also differ according to morphology (e.g.
Alexandria vz Port Said in Egypt, [12]).

## Adaptation

The limit of active defence will be decided by the social and
political acceptability of increasing hazard risks and the cost
of direct protection.  Beyond that limit, social adaptation and
changes in land use would be preferable, such as moving industrial plants inland; reverting sub-sealevel reclaimed lands to
their former lagoonal state (lagoons and marshes could act as
buffer zones between the open sea and higher land, as well as
nature reserves); modification of farming practices; development
of salt tolerant plants; a more sensible approach towards the
management of tourist beach facilities, such as to establish
setback lines and the adoption of more 'open space' land use,
with less urbanized settlements, perhaps following the model of
the Languedoc-Camargue coast in France [8].

In the Mediterranean region the degree of the adaptation
would vary; clearly it would be a question of local situations
and of economic, and especially political expediency (assuming
that areas that look suitable for adaptation today, would still
so in 20-30 years time).

Nevertheless the complexity and large scale nature of
adaptation would entail technical, financial and administrative
problems and would call for an integrated approach at a regional
level rather than ad-hoc local interventions.

## Abandonment/planned retreat

If deployment would be excessively costly and general disruption
resulting from trying to hold back the sea, substantial, abandonment may become unavoidable. It could be imposed anyway by economic stringency: the excessive burden imposed by escalating
costs of shifting coastal infrastructure, reorganizing beach
resorts, moving population, on state and local government budgets,
perhaps taking up a large percentage of the income generated by
coastal economies.

Conscious retreat would probably make sense only in sparsely
populated areas for example those involving reclaimed low lands
with extensive agriculture (e.g. margins of Po, Ceyhan, Medjerda
deltas). However, if such areas become more densely populated
(as a result of incorrect planning), displacing populations
would involve difficult political decisions.

Therefore, in the Mediterranean abandonment of coastal lands
would probably be confined to isolated localities and it would
happen only under forced circumstances.

## Policies

It is certainly difficult to determine, now, what policies of responses ought to be adopted, where, or if any, at what stage, given the Mediterranean region's diverse degree of vulnerability, according to morphology, exposure to waves, present and projected levels of occupation,

Policies may in fact not be the result of a deliberate choice, but would be imposed by the scale and frequency of disastrous natural events, or by the processes of market economy, as at many places public and private investments might gradually become economically un-sustainable, because increased physical damages make maintenance too costly (e.g. polders infrastructures, leisure harbours, power stations by the sea, etc.); and/or, especially, by priorities, such as:

    a) other consequences of climate change demanding society's attention (e.g. the impacts on energy consumption, water resources and agriculture);

    b) local coastal interests not being the same as national or international public interest;

    c) regional, provincial or other local authorities covering predominantly inland territories, with short coasts of no, or of relatively minor economic importance, may not be willing to share in broader coastal planning [31].

The best policy option for Mediterranean coasts in the forthcoming two decades appears to be, in fact, that of integrated coastal zone planning and management (ICZPM). This is the approach that would best deal with both present and future, anthropic and climate-related problems.

A symptom-conditioned or socio-economic sectorial approach cannot be the basis for rational development of areas that are characterized by constantly changing physical conditions, by competition for resources and space, by the fact that development at one place alters conditions elsewhere [39]. ICZPM requires a multidisciplinary approach, with evaluation and integration of all technological, social, economic, cultural, ecological and institutional elements. It is a long-term, continous and cyclic process, within which mechanism of response are guided by changes.

Present environment-related objectives are, for instance sound planning to reduce beach congestion and stress, to obtain more efficient relations between tourist centers and their hinterlands, to estimate eventual loss of revenues.

The evaluation of climatic change consequences should be a similar process, because of the need to emphasize levels of events probability, of alternative responses. Solutions are to work in multiple ways (for example, the evaluation of the extent of social disruption as a consequence of inland retreat).

The concept of changing climate should be introduced into the coastal planning process. In the time framework of the present SLR-trend scenario, considering that major coastal management projects take up to 30 years to be achieved (the time required for technical formulation, planning, political and financial decisioning, legal instrumentation), the coastal management time horizon would be 30-50 years, with adaptation necessary, perhaps before 2050. If the predicted trend is, instead of +60cm by the late 2000's, implementation of SLR-related CZM and adaptive options by 2050-60, would probably require action to start about 2020-30, if not earlier.

Two examples of how development could be twarthed by inadequate appreciation of the consequences of coastal zone mismanagement, or by insufficient potential for sustainable development, in the context of possible future climatic change, are the northern Nile delta and the SE Adriatic coast of Albania, areas of respectively medium-moderate, and of very low present development.

In the Nile delta, the nature and extent of SLR and climate change impact would depend largely on the degree of coastal development during the next 2-3 decades, as population pressure will increase the state of occupation of the coastal strip under 1m. Schemes of integrated lagoon management are urgently needed to evaluate and harmonize a number of very contrasting projects of wetland use that are envisaged for the near future: expansion of aquaculture versus that of agriculture, use of lakes for freshwater storage, the building of roads and new towns on sand barriers by the shore [66]. Some of the proposed projects are in high risk areas, because of the serious danger of shore retreat and flooding. Econo-mic and social impacts could therefore be very negative, if planning is mismanaged.

In Albania tourist exploitation of a still undeveloped coast (which contains largely unspoiled natural ecosystems) may result in duplication of all the negative side effects of anthropic actions in the NW Adriatic coast in Italy. At least, resorts of 'open type' should be located with full knowledge of shoreline dynamics and of trends of decadal changes [52]. Integrated planning is necessary also in regard to the use of lagoons *versus* the needs of agriculture, the location of industries in relation to likely pollution.

## THE WAY FORWARD

It stands to reason that many of the adverse economic impacts of sea-level rise (whether periodic, exceptional or long-term) might be avoided if timely anticipatory actions are taken. The guiding principle should undoubtedly be that of preventing sea-level rise related disasters, of minimising the adverse effects of development, of fostering the goal of sustainable coastal zone development and management, and of the conservation of natural resources, not least, in view of the consequences of potential major climatic changes.

In order to integrate natural and man-made systems, what is needed is a sound environmental data base [2,49]. The following is a list of topics for investigation, gleaned from a number of impact assessment studies [e.g. 29,30,45,49,75].

## Coastal zone inventory and monitoring

1. Sea-level changes monitoring.  2. Periodicity of climate-induced natural events (low-frequency vz high frequency variability, likely period of recurrence); improved prediction/warning of extreme weather events and of flooding. 3. Knowledge of future shorelines, based on trends of coastal changes from historical data from sequential topographic maps, aerial photography satellite images). 4. Inventory and mapping (by any suitable technique) of anthropic shoreline activities. 5. Definition of level of vulnerability, identification  and mapping of the high-risk areas (incl. size of population, activities affected).  6. Studies of local coastal dynamics and sedimentary budgets with higher SL (esp. in function of artificial beach nourishment).  7 Identification of the coastal stretches that are rising or subsiding (prediction of future coastal earth movements [17]). 8. Groundwater flows and saline seepages. 9. Collection of causal and quantitative information on the interaction of the biotic and abiotic environments in coastal wetlands and the shallow marine zone.

## Modelling

Development and/or application of quantitative models to assess the interrelations of the listed coastal variables in relation to different degrees of SLR risk, as well as to scenarios of demographic and economic development at given time scales (examples are the existing ISOS, CISCA, WSBM, LIFE models [22,88]).

## Coastal zone planning and management

1. Appraisal of the requirement of each planned economic activity (beach tourism, marinas, industries, commercial fishing, aquaculture, land transport, ports, agriculture, offshore oil/gas exploitation; in terms of commercial viability, and of the intersectorial feedbacks.

2.  Environmental impact assessment of each coastal use. Analysis of ecological-economic impacts in terms of multiple stress.

3.  In coastal zones yet to be developed: assessment of the inte-gration of economic activities, especially for a rational management of beach tourism.

4.  Cost benefit analyses: research into the approach  vz time scales, estimation of costs vz options scenarios; cost of adaptation viewed on a long-term economic base, and within national perspectives.

5.  Evaluation of how to control coastal development (e.g. land reclamation; industries; excessive groundwater exploitation; pollution, energy and waste discharge; the dumping/

storage of toxic and radioactive wastes in lowlands liable to
inundation), with the objec-tive of land-use regulation and
zoning according to degree of SLR risk, and to prevent increased
future vulnerability; of how to reconcile nature protection
with necessary development.

6. How to match management organization and instruments
with the scale of the pro-blems: coordination of coastal
policies, delimitation of the respective roles of the public and
private sectors; administrative and legal framework of decision
making process; creation of national institutions responsible
for advance strategies towards climatic hazards.

**Promotion of awareness**
Scientific appraisal on one hand, and public and official awareness, on the other, need to be brought more into line.

1. General awareness is needed that coastline instability
(i.e. erosion and accretion) is natural. The coast and sea-level
are shifting, not fixed entities. Vice-versa an anthropically
fixed coastline is not natural and could persist only at high
cost.

2. There is a need for national response strategies toward
natural hazards (whatever the cause) and to include major
climatic changes sce-narios into the general planning and
policy-making process.

3. Awareness that planning and controlling coastal development requires administrative and legal modifications and above
all, political will. Management instruments, at institution
level, need to be streamlined, and national organizations are
necessary (or need strengthening) to provide prescient responses
to natural hazards, instead of the common approach of ad hoc
solutions.

4. Finally, awareness of the need to develop a common
unders-tanding, between the various professional groups that deal
with coasts: engineers, natural scientists, economists, archi-
tects, lawyers, administrators; who normally tend to talk quite
different languages.

## CONCLUSIONS

Sea-level in the Mediterranean is areally and temporally very
variable, due to geological factors and to diverse tidal and
meteorological forcing. Storm surges in association with spring
tides, have/ are causing considerable damages to settlements and
land uses, especially in the most vulnerable areas: low coasts
that have become the focus of activities, subsiding deltas,
sediment-starved beaches. The impacts of exceptional weather
events are exacerbated by the present state of degradation of
many coastal and shallow marine areas.

There is a fair likelihood, although no concrete evidence
yet, of substantial climatic changes during the next century,

with an attendant acceleration of sea-level rise. Given the variability of sea-level rise factors, it is impossible to predict future increments at specific localities; however, if changes materialize according to probabililistic scenarios,, the consequences would include a large amount of damage, of disruption of economic activities, and growing expenditures for adaptation.

In view of rising anthropic pressure on coastal zones and of their exacerbation of environmental problems, main recommended action at this stage is to create a basis for disaster avoidance and to enhance preparedness for future events by implementing integrated coastal zone management, both in developed and developing coasts. Its aim should be sustainable development and the protection of natural resources. Basic starting points are the assembly of exhaustive data bases, for all coastal parameters (physical, socio-economic), the monitoring of many, and the streamlining of institutional and legal instruments.

## REFERENCES

1. Baric, A., Impacts of climate change on the socio-economic structure and activities in the Mediterranean region. In Changing Climate and the Coast, J.G.Titus (ed.), U.S. EPA, 1990, **2**, 127-138.

2. Baric, A. and Gasparovic, F., Implications of climatic change on the socio-economic activities in the Mediterranean coastal zones. In Climate Change and the Mediterranean, L.Jeftic,J.D.Milliman and G.Sestini (eds.), E.Arnold, London, 1992, pp. 131-176.

3. Bird, E.C.F. and Paskoff, R., Geomorphological problems on the Mediterranean coasts. In Coastal Problems in the Mediterranean Sea, E.C.Bird and P.Fabbri (eds.), I.G.U. Commiss.Coastal Environments, Bologna, 1983, pp. 7-17.

4. Bird, E.C.F. and Schwartz, M.L. (eds.), The World's Coastlines, Van Nostrand Reinhold, New York, 1985, pp. 385-544.

5. Brambati, A., Erosione e difesa delle spiagge adriatiche. Boll. Oceanologia Teorica Applicata, 1984, **2**, pp. 91-115.

6. Carbognin L., Gatto, P. and Mozzi, G., La riduzione altimetrica del territorio veneziano. Ist. Veneto Lett. Sci. Arti, Rapporti e Studi, 1981, **VIII**, pp. 55-83.

7. Cencini, C., Marchi, M., Torresani, S. and Varani, L., The impact of tourism on Italian deltaic coastlands: four case studies. Ocean and Shore Management, 1988, **11**, pp. 353-374.

8. Corre, J.J., Implications des changements climatiques, etude de cas: Le Golfe de Lion, France. In Climate Change and the Mediterranean. L.Jeftic, J.D.Milliman and G.Sestini (eds.), E.Arnold, London, 1992, pp.330-428. Also: Implications of climatic change for the Gulf of Lion. UNEP, OCA/PAC, Nairobi, 1989 Report WG.2/5.

9. Dal Cin, R., I litorali del delta del Po e alle foci dell' Adige e del Brenta, caratteri tessiturali e dispersione dei sedimenti, cause dell' arretramento e previsioni sull' evoluzione futura. Boll.Soc.Geol.Ital., 1983, 102, pp. 9-56.

10. Delft Hydraulics, Implications of relative sea-level rise on the development of the lower Nile delta, Egypt. Pilot study for a quantitative approach. Delft Hydraulics, 1991, Rept. H 927.

12. El Sayed, M.K., Implications of relative sea level rise on Alexandria, In Impacts of Sea Level Rise on Cities and Regions, R.Frassetto (ed.), Marsilio Editori, Venice, 1991, pp. 183-189.

13. Emery, K.O., Aubrey, D.G. and Goldsmith, V., Coastal neotectonics of the Mediterranean from tide-gauge levels. Marine Geology, 1988, 81, pp. 41-52.

14. Erol, O., Impacts of sea level rise on Turkey. In Changing Climate and the Coast, J.G. Titus (ed.), U.S. EPA, Washington D.C., 1990, 2, pp.183-200.

15. Fabbri P., Introduction. In Recreational uses of coastal areas, P. Fabbri (ed.), Kluwer Academic Publishers, 1990, pp.vii-xviii

16. Fairbridge, R.W., Crescendo events in sea level change. J.Coastal Res., 1989, 5, pp. ii-vi.

17. Flemming, N.C., Predictions of relative coastal-sea level change in the Mediterranean, based on archeological, historical and tide gauge data. In Climate Change and the Mediterranean, L.Jeftic, J.D. Milliman and Sestini, G. (eds.), E.Arnold, London, 1992, pp. 249-283.

18. Flemming, N.C. and Woodworth, P.L., Monthly mean sea levels in Greece during 1969-83 compared to vertical land movements measured over different tipe scale. Tectonophysics, 1988, 148, pp. 59-62.

19. Frihy, O.E.,Nile delta shoreline changes: aerial photographic study of a 28-year period, J.Coastal Research, 1988, 4, pp.497-506.

20. Gacic M., Hopkins T.S. and Lascaratos A., Aspects of the response of the Mediterranean Sea to long term trends in atmospheric forcing. In Climate Change and the Mediterranean, L.Jeftic, J.D.Milliman and G.Sestini (eds.), E.Arnold, London, 1992, pp. 60-85.

21. Gatto, P. and Carbognin, L., The lagoon of Venice, natural environmental trend and man-induced modifications. Hydrological Sciences Bull., 1981, 26, pp. 379-39.

22. Georgas, D. and Perissoratis, C., Implications of future climatic changes on the inner Thermaikos Gulf. In Climate Change and the Mediterranean, L.Jeftic, J.D.Milliman and G.Sestini (eds.), E.Arnold, London, 1992, pp. 496-534. Also: UNEP, OCA/PAC, Nairobi, 1989, Rept WG 2/9.

23. Gibbs, M.J., Economic analysis of sea level rise: methods and results. In: Greenhouse Effect and Sea Level Rise, M.C.Barth and Titus J.C. (eds.), Van Nostrand Reinhold Co., 1984, pp. 215-250.

24. Gornitz, V., and Lebedeff, S., Global sea level changes during the past century. In Sea level fluctuations and coastal volution, Nummedal D., O.E. Pilkey and J.D.Howard (eds.), S.E.P.M., Tulsa, 1987, Spec. Publ., 41, pp.3-16.

25. Hekstra, G.P., Will climatic changes flood the Netherlands? Effects on agriculture, land-use and wellbeing. Ambio, 1986, 17, pp.316-326.

26. Hendricks, J.C.F., Remery, F.J., de Ronde, J.G., de Swart P.F. and Vrijling, J.K., Effects of rise of sea-water level on maritime structures in the Netherlands. Rijkswaterstaat, 1989 PLANC Bull., 66

27. Hollis, G.E., The modelling and management of the internationally important wetland at Garaet El Ichkeul, Tunisia. IWRB Special Publ., 4, 1986, 121 pp.

28. Houghton, J.T., Jenkins, G.J. and Ephraums, J.J. (eds), Climate Change, the IPCC Scientific Assessment. University of Cambridge Press, 1990.

29. Jeftic L., Milliman, J.D. and G.Sestini (eds.), Climate Change and the Mediterranean. E.Arnold, London, 1992, 685 pp.

30. Jelgersma, S. and Sestini, G., Implications of a future rise in sea level on the coastal lowlands of the Mediterranean. In Climate Change and the Mediterranean, L.Jeftic, J.D.Milliman and G.Sestini (eds.), E.Arnold, London, 1992, pp.285-305.

31. Jolliffe, I.P. and Patman, C.R., The Coastal zone: the challenge. J. Coast. Management, 1985, 1, pp. 3-36.

32. Komar, P.D. and Enfield, D.B., Short-term sea-level changes and coastal erosion In Sea level fluctuations and coastal evolution, D.Nummedal, O.E. Pilkey and J.D.Howard (eds.), S.E.P.M., Tulsa, 1987, Spec. Publ., 41, pp. 17-28.

33. Lamb, C., Climate in the last thousand years: natural climatic fluctuation and change. In The Climate of Europe: past, present and future. H.Flohn and R.Fantechi (eds.), D.Reidel Publishing Company, 1988, pp. 25-64.

34. Lascaratos, A., Interannual variations of sea level and their relation to other oceanographic parameters. Boll. Oceanogr. Teorica Applicata, 1989, VII, pp. 317-321.

35. Lascaratos, A., Gacic, M. and Oguz, T., Climatic and sea level variability in the northeastern Mediterranean and Black Sea. Proceed. ICSEM Meeting, October 1990, Perpignan, 1990.

36. Leatherman, S.P., Modelling shore response to sea level rise on sedimentary coasts. Progress in Physical Geography, 1990, 14, pp.447-464.

37. Laurman, J.A., Global warming and credibility of science. Climatic Change, 18, pp.107-109.

38. Lough, J.M., Wigley, T.M.L. and Palutikov, J.P., Climate and climate impact scenarios for Europe in a warmer world. J. Climate and Applied Meteor., 1983, 22, pp. 1673-1684.

39. M.A.P.- UNEP, Methodology for assessment and mitigation of impacts of expected climate changes and sea level rise within the process of integrated planning and management of coastal zones. Priority Actions Programme, Regional Activity Center, Split, 1991, UNEP(OCA)-WG.12

40. Marino, M., The impacts of climatic changes on the Ebro delta. In Climate Change and the Mediterranean, L.Jeftic, J.D.Milliman, pp. 306-329. Also: UNEP, OCA/PAC, Nairobi, 1989, Rept. WG/2/3.

41. Maroukian, H., Implications of sea level rise for Greece. In Changing Climate and the Coast, J.G.Titus (ed.), U.S. EPA, Washington D.C., 1990, 2, pp.161-182.

42. Maul, G., Implications of climatic changes in the wider Caribbean Region. Preliminary conclusions of Task Team of experts. Caribbean Environment Programme, UNEP-CEP, Kingston, Jamaica, 1989, Techn.Rept., 3.

43. Mazzarella, A. and Palumbo, A., Long period variations of mean sea level in the Mediterranean Area. Boll.Oceanol. Teor.Applicata, 1989, 6, pp. 253-259.

44. Mazzarella, A. and Palumbo, A., Effect of sea-level time variations on the occurrence of extreme storm-surges: an application to the north Adriatic Sea, Boll.Ocean..Teor. Applicata, 1991, 9, pp. 33-38.

45. Mehta, A.J. and Cushman, R.M., Workshop on Sea Level Rise and Coastal Processes, U.S. Dept.Energy, Office of Energy Research, 1989, DOE/NBB-0086, 286 pp.

46. Miller, A., The Mediterranean Sea. Physical Aspects. In Estuaries and Enclosed Seas, B.H.Ketcheem (ed.), Ecosystems of the World, 26, Elsevier Sci.Publ.Co., 1983, New York, pp. 219-238.

47. Milliman, J.D., Sea level response to climate change and tectonics. In Climate Change and the Mediterranean, L.Jeftic, J.D.Milliman and G.Sestini (eds.), E.Arnold, London 1992, pp. 47-59.

48. Milliman J.D., Broadus, J.M. and Gable, F., Environmental and economic implications of rising sea-level and subsiding deltas: the Nile and Bengal examples. Ambio, 1989, 18, pp. 340-345.

49. Misdorp, R., Dronker, J. and Spradley, J.R. (eds.), Strategies for adaption to sea level rise. IPCC Report of the Coastal Zone Management Subgroup. The Netherlands, Ministry of Transport and Tidal Waters, Rijkswaterstaat, Tidal Waters Division, 1990, 121 pp.

50. Mosetti, F., Sea-level variations and related hypotheses. Boll. Oceanol. Teorica Applicata, 1989, VII, pp.273-284.

51. Nafas, M.G., Fanos, A.M. and Elganainy, M.., Characteristics of waves off the Mediterranean coast of Egypt, J. Coastal Res., 1991, 7, pp. 665-676.

52. Oueslati, A., Coastal morphology and sea level rise consequences in Tunisia. In Changing Climate and the Coast, J.G.Titus (ed.), U.S. EPA, Washington, D.C., 1990, 2, pp. 211-224.

53. Pano, N. and Sestini, G., Coastal changes and coastal zone management in Albania. 1992, in preparation.

54. Pilkey, O., A time to look back at beach replenishment. Journ. Coastal Research, 1990, 6, pp iii-vi.

55. Pirazzoli, P.A., Sea level changes in the Mediterranean. In Sea Level Change, M.J.Tooley and S.Jelgersma (eds.), Basil Blackwell, Oxford, 1987, pp.152-1810.

56. Pirazzoli, P.A., Recent sea-level changes and related engineering problems in the Lagoon of Venice, Italy. Progr. in Oceanogr., 1987, 18, pp. 323-346.

57. Pirazzoli, P.A., Sea level changes and crustal movements in the Hellenic arc (Greece). The contribution of archeological and historical data. In Archaeology of coastal changes, A.Raban (ed.), BAR Int.Sr. 404, 1988, Publ. 2, pp. 157-183.

58. Pirazzoli, P.A., Present and near-future global sea level changes. Global Planetary Change, 1989, 1, pp. 241-258.

59. Pirazzoli, P.A., World Atlas of Sea Level Changes. Elsevier, Amsterdam, 1991, 300 pp.

60. Ramanathan, V., The greenhouse theory of climatic change: a test by an advertent global exchange. Science, 1988, 240, pp. 293-299.

61. Reed, D.J., The impact of sea level rise on salt marshes. Progress in Physical Geography, 1991, 14, pp. 465-4   .

62. Sestini, G., Impacts of global climate change in the Mediterranean region: responses and policy options. In Changing Climate and the Coast, J.G. Titus (ed.), U.S. EPA, Washington D.C., 1990, 2, pp. 115-126.

63. Sestini G., Implications of climatic changes for the Po Delta and the Venice Lagoon. In Climate Change and the Mediterranean, L.Jeftic, J.D.Milliman and G.Sestini (eds.), E.Arnold, London, 1992, 429-495. Also: UNEP, OCA/PAC, Nairobi, 1989, Rept WG 2/11.

64. Sestini, G., Implications of climatic changes for the Nile Delta. In Climate Change and the Mediterranean, L.Jeftic, J.D.Milliman and G.Sestini (eds.), E.Arnold, London, 1992, 535-600 . Also: UNEP OCA/PAC, Nairobi,1989, Rept.WG 2/3.

65. Sestini, G., The impact of climatic change on two deltaic lowlands in the Eastern Mediterranean. In The Impact of Sea Level Changes on European Coastal Lowlands, J.M.Tooley, S.Jelgersma (eds.), Inst.Br. Geogr. Spec.Publ., Basil Blackwell, Oxford, 1992, in press.

66. Sestini, G., The consequences of climatic change for the Mediterranean lagoons. In Managing Mediterranean Wetlands and Their Birds for the Year 2000 and beyhond. Proc. Grado Febr.1990 Conf., IWRB Report, 1992, in press.

67. Sestini, G., The effects of sea-level changes in the Mediterranean region. A review of the main issues. UNEP-MAP Techn. Reptorts Series, Athens, 1992.

68. Sestini, G., Jeftic, L. and Milliman, J.D., Implications of climatic changes in the Mediterranean, an overview. UNEP, Regional Seas Reports and Studies, Nairobi,1990, 103, 49 pp.

69. Sharaf El Din, S.H. and Moursi, Z.A., Tide and storm surges on the Egyptian Mediterrannean coast. Rapp.Comm.Intern. Mer. Med., 1977, 24, pp. 33-37.

70. Sharaf el Din, S.H., Ahmed, K.M., Fanos, A.M. and Ibrahim, A.M., Estreme sea-level values on the Egyptian Mediterranean coast for the next 50 years. In Proceedings Intern. Seminar on Climatic Fluctuations and Water Management, Cairo 1990, Paper II-6.

71. Stanley, D.J., Recent subsidence and northeastern tilting of the Nile delta. Marine Geology, 1990, 94, pp. 147-154.

72. Thompson, R.D., Short-term climatic change: evidence, causes, environmental consequences and strategies for action. Progress in Physical Geography, 1989, 13, pp. 315-347.

73. Titus, J.G., The causes and effects of sea level rise. In Impacts of Sea level rise on Society, Wind, H.G. (ed.), A.A. Balkema, Rotterdam-Brookfield, 1987, pp. 104-125.

74. Titus, J. G. (ed.), Greenhouse effect, sea level rise and coastal wetlands. Washington, U.S. EPA, 1988, Rept. 230-05-86-113.

75. Titus, J.G. (ed.), Changing Climate and the Coast. Vol.1, Adaptive responses and their economic, enviornmental and institutional implications; vol.2, Western Africa, the Americas, the Mediterranean basin, and the rest of Europe, U.S. EPA, Washington D.C., 1990, 508 pp.

76. Titus, J G and Green, M.S., An overview of the nationwide impacts of sea level rise. In The potential effects of Global climatic Change on the United States. Appendix B, Sea Level Rise, J.B.Smith and D.A.Tirpak (eds.), U.S. EPA, Washington D.C., 1989, 230-05-89-052, pp.(5)1-54.

77. UNEP, The Blue Plan, Futures of the Mediterranean. Executive summary and suggestions for action. Blue Plan, Sophia Antipolis, 1988, 641 pp.

78. UNEP, State of the Mediterranean Marine Environment. MAP Techn. Rpts Series, 28, Athens, 1989.

79. Vellinga, P., Beach and dune erosion during storm surges. Delft Hydraulics Laboratory, 1986, Publ. No 372.

80. Walker, H.J. (ed), Artificial structures and shorelines, New York Kluver Academic Publishers, New York, 1988.

81. Warrick, R.A. and Farmer, G., The greenhouse effect, climatic change and rising sea level: implications for development. Trans. Inst.Br. Geogr., 1989, n.s. **15**, 5-20.

82. Warrick, R.A. and Oerlemans, J., Sea level rise. In Climate Change, the IPCC Scientific Assessment, J.T.Houghton, G.J.Jenkins and J.J. Ephraums (eds), University of Cambridge Press, 1990, pp. 261-285.

83. Warrick, R.A. and Wigley, T.M.L. (eds.), Climate and sea level change. Observations projections and implications. Cambridge Univ Press, Cambridge, 1991, in press.

84. Wigley, T.M.L., Future climate of the Mediterranean basin, with particular emphasis on changes in precipitation. In Climate Change and the Mediterranean, L.Jeftic, J.D.Milliman and G.Sestini (eds.), E.Arnold, London, 1992, pp. 17-45.

85. Wigley, T.M.L., The effect of changing climate on the frequency of absolute extreme events. Climate Monitor, 1989, **17**, pp. 44-55.

86. Wigley, T.M.L. and Raper S.C.B., Thermal expansion of sea water associated with global warming. Nature, 1987, **330**, pp. 127-131.

87. Wiin-Nielsen , A., The Greenhouse Effect. A review of data and model studies. In Impacts of Sea Level Rise on Cities and Regions, R.Frassetto (ed.), Marsilio Editori, Venice, 1991, pp. 25-32.

88. Wind, H.G. (ed.), Impacts of Sea level rise on Society. A.A. Balkema, Rotterdam-Brookfield, 1987, 191 pp.

89. WMO/UNEP, Climatic Change. The IPCC Impacts Assessment. Working Group II. Chapter 6: World Oceans and Coastal Zones. M.J. McG. Tegart, G.W.Sheldon and D.C.Griffith (eds.), Australian Gvmt Publishing Service, Camberra 1990.

90. Zavatarelli, M., Potential impact of the greenhouse effect on the Mediterranean Sea, an overview. Intern.Inst.of Applied Systems Analysis, Laxenburg, Working Papers, XWP-88-76, 1987, 28 pp.

91. Zunica, M., Beach behaviour and defences along the Lido di Jesolo, Gulf of Venice, Italy. J.Coast. Research, 1990, **6**, pp.709-719

# COASTAL ZONE MANAGEMENT
## EXPERIENCE OF MEDITERRANEAN ACTION PLAN - UNEP

ARSEN PAVASOVIC
Director
Regional Activity Centre for the Priority Actions Programme
Kraj Sv. Ivana 11, 58000 Split, Croatia

## ABSTRACT

The Mediterranean Action Plan (MAP) is a specific way of cooperation of all Mediterranean countries and the CEC in the field of the protection of the Mediterranean against pollution. MAP is implemented as component of UNEP - Regional Seas Programme, on the legal basis of the Barcelona Convention. MAP is a comprehensive set of harmonized activities aimed at monitoring and research of the state of pollution of the Mediterranean, measures for the protection against pollution, and finally at strengthening, or creating conditions for the application of the concept of sustainable development through the process of integrated coastal zone management.

A survey of most important marine and coastal resources is also presented, as well as the influence of uncontrolled development processes and pollution to which those resources are exposed. The term "coastal zone" is defined. Since those zones are, as a rule, the richest development resources of each country, and due to the complexity and a large number of inter-relations occurring in those zones, a new way of thinking and a new approach must be applied in order to achieve their rational development and protection. This particularly refers to the need for relevant national bodies to have authority over the process of development and utilization of coastal zone resources.

The UNEP/MAP concept of integrated coastal zone management and the principles of sustainable development are also presented. A common methodological framework for the integrated planning process, applicable in the Mediterranean developing countries, is described, as developed by MAP/PAP. Description is given of the results of MAP in the development of the methodology and application of environment-development scenarios and other tools and techniques of integrated coastal zone planning and management. The need is pointed out of including immediately the mitigation of impacts of expected climate changes into the process of integrated coastal zone management.

Finally, the paper presents the concept, objectives and contents of MAP Coastal Area Management Programmes (MAP CAMPs) implemented, in cooperation with the national authorities, in the Bay of Izmir, the island of Rhodes, the Kastela Bay, and the Syrian coast. The basic objectives of those programmes are the completion of environmental knowledge on the relevant coastal areas, creation of conditions for the application of the process of ICZM and its tools and techniques, and training and upgrading of national and local institutions and experts on these subjects.

Although those programmes are still in the phase of implementation, the achieved experience confirms the validity of the selected approach, and the need of a careful preparation of institutional arrangements. Finally, some basic results and outputs are presented.


## PREFACE

This paper has been prepared to be presented at the Semi-Enclosed Seas Conference (Genoa, February 1992). Due to its limited length, the paper is a condensed information on MAP/UNEP activities and experience relative to coastal zone management in the Mediterranean region. Any further information can be obtained consulting the bibliography or, upon request, from the author.

The opinions stated in the paper are personal opinions of the author and do not necessarily reflect the opinions of UNEP, MAP and PAP/RAC.


## SOME BASIC FACTS ON THE MEDITERRANEAN ACTION PLAN - UNITED NATIONS ENVIRONMENT PROGRAMME (UNEP)

Within the UNEP Regional Seas Programme (RSP), a total of 11 regional action plans are implemented, of which the Mediterranean Action Plan (MAP) is the leading and most developed one. The legal basis of MAP is the Convention on the Protection of the Mediterranean Sea against Pollution, and its related Protocols, signed in 1976 in Barcelona (the "Barcelona" Convention). That Convention was signed by all Mediterranean countries and the Commission of European Communities (CEC). MAP is a comprehensive programme of cooperation oriented at research into the causes of and protection against pollution of the Mediterranean Sea and its coastal zones. MAP consists of 4 major components: (a) the scientific component dealing with research and monitoring of the state of pollution of the Mediterranean Sea; (b) the integrated planning and management component oriented at prospective studies of environment-development interrelations, and at the implementation of the process of integrated planning and management; (c) the legal component - application of the Convention and its Protocols; and (d) institutional arrangements. A simplified organizational scheme of MAP is given in Fig. 1.

The scientific component (MEDPOL) deals with the assessment of the state of pollution, identification of pollutants and trends in the levels of pollution, proposals of methods of pollution control and management, and provision of relevant inputs for coastal zone management. In 1983 MEDPOL started preparation and, later, implementation of the Protocol for the Protection of the Mediterranean Sea against Pollution from Land-Based Sources. MEDPOL is coordinated by the MAP Coordinating Unit in Athens, and implemented through national MEDPOL programmes.

The socio-economic or integrated planning component consists of two programmes: the Blue Plan and the Priority Actions Programme.
The Blue Plan is a programme of prospective studies, using scenario methods related to development-environment interrelations with the years 2000 and 2025 as time horizons, with the objective of providing inputs for environmentally sound development, planning and management in the region. The Blue Plan is implemented and coordinated by the Blue Plan Regional Activity Centre (BP/RAC) located at Sophia Antipolis, France.
The Priority Actions Programme (PAP), is implemented by the Regional Activity Centre (PAP/RAC) in Split, Republic of Croatia. It is an action-oriented programme aimed at carrying out practical actions, and contributing - through exchange of available knowledge and experience among the Mediterranean countries - to the protection and enhancement of the Mediterranean environment and to the strengthening of national and local capacities for the implementation of the process of integrated planning and management of coastal zones. The PAP Workplan comprises 10 individual priority actions relative to: coastal zone planning and management, historic settlements, seismic risk, water resources management, urban solid and liquid wastes, soil protection, tourism, aquaculture, renewable sources of energy, and environmental impact assessment.

The legal component of MAP is oriented to the implementation of the legal aspects of the Barcelona Convention and its related protocols. The following protocols were adopted by the Contracting Parties: (a) for the protection against dumping from the ships and aircrafts (adopted in 1976 by 18 countries and CEC); (b) for combating pollution by oil and other harmful substances in case of emergency (1976, 18 countries and CEC); (c) for the protection from land-based sources of pollution (1980, 15 countries and

113

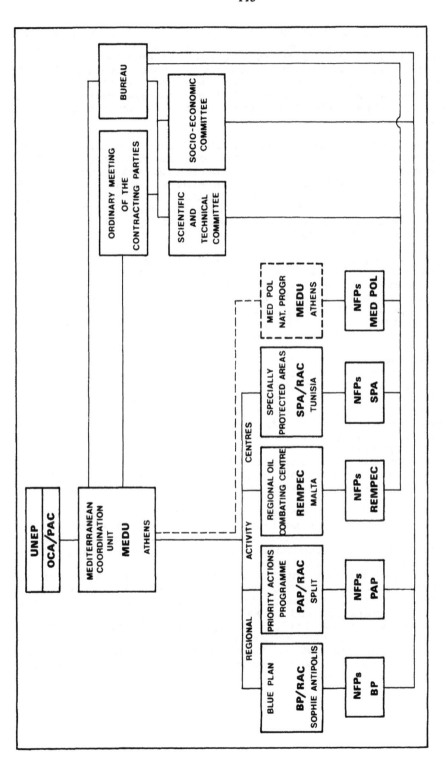

**Figure 1**
**MAP-ORGANIZATIONAL CHART**

CEC); and (d) concerning Mediterranean **specially protected areas** (1982, 16 countries and CEC). The activities related to the protocol on pollution by oil are implemented through the **Regional Marine Pollution Emergency Response** Centre **(REMPEC)** situated in Malta, while those relative to the specially protected areas are implemented by the **Regional Activity Centre for Specially Protected Areas (SPA/RAC)** situated in Tunis.

The institutional component of MAP includes institutional and financial arrangements. UNEP acts as the secretariat of MAP and coordinates the implementation of the Plan. On that basis the MAP Coordinating Unit in Athens, as a part of UNEP, coordinates the work of all programmes, implements MEDPOL and all activities related to the international cooperation of MAP. Individual programmes are performed by and through the Regional Activity Centres.

Financing of MAP is secured through contributions by the Contracting Parties to the Mediterranean Trust Fund administrated by UNEP. A part of the finances is provided by the countries hosting the RACs. A considerable financial support is provided by the contributions other than budget, by the individual contracting parties or other donors (such as The World Bank and the European Investment Bank).

Recent developments led to some new **MAP** activities: identification and protection of 100 historic sites of common Mediterranean interest (implemented by a new Project Activity Centre in Marseille), programmes for the protection of endangered species, and for the assessment of impacts of climate changes, while the Protocol on Protection against Pollution Caused by Exploration and Exploitation of the Sea Bed is in preparation.

After the decision to gradually refocus the entire programme on coastal zone management, MAP started in 1989 a number of pilot programmes in selected Mediterranean coastal zones.

Although MAP is implemented through official governmental (national) authorized bodies, it developed, especially since the refocusing on coastal zone management, such a way of cooperation which includes sub-national (regional, local) authorities and institutions.

An ever growing importance in the implementation of MAP has been gained by the CEC whose environmental programme, and particularly the implementation of the Charter of Nicosia of 1990, is closely related to MAP.

A number of specialized UN agencies and bodies have been taking part in the MAP implementation, such as UNDP, WHO, FAO, UNESCO, UNCHS and UNDRO. That secures the application and integration of the best and updated specialized knowledge, enabling, at the same time, a rational use of funds.

## MARINE RESOURCES AND COASTAL ZONES

All constituents of the marine environment that could be used in the process of development, as well as all elements of permanent value making the natural environment for such a development can be considered as marine resources. Those are: coastlines, deltas and estuaries, lagoons, bays, coastal wetlands and marshes, natural and archaeological marine parks and other submarine areas of particular value, off-shore mineral resources and minerals dissolved in the sea, and the marine animal and vegetal life. Specific values of scientific, technical, cultural and historic nature of both coastal sea and coastal zones also have to be added. Finally, it is necessary to mention industrial and other economic activities situated in the coastal zones (marine transport, harbours, tourism, agriculture, recreation).

For the majority of coastal states, and especially for the island ones, the marine and coastal resources are the most important development resources. It has to be pointed out that today, coastal nations make more than 80% of the members of the UN, of which 40 are island nations (in the Mediterranean those are Cyprus and Malta).

Besides the purely qualitative evaluation of the marine and coastal resources of each country, it is possible to show a monetary value of coastal resource production and activities, export earnings, employment, and a number of non-market ones.

Due to a strong, often uncontrolled development of coastal zones, urban expansion, and high rate of population growth, the marine and coastal resources are under permanent pressure. Pollution processes in many areas have already caused serious, often irreversible damages. The principal sources of pollution are mostly the same for all regional seas. Some of them could be illustrated at the example of the Mediterranean. Of 360 million inhabitants of the Mediterranean coastal states, 38% live in the coastal zones. Larger part of industrial wastewaters and about 70% of urban wastewater of the Mediterranean towns are discharged into the sea treated insufficiently, inadequately or not at all. 40% of the total pollution load is borne by 5 big rivers. An ever increasing amount of transboundary pollution also has to be added, as well as diffuse sources of pollution. The consequences of such pollution are, among others, pollution of surface and ground fresh waters, soil degradation, rupture of ecosystems, irreversible loss of national resources, and a number of negative socio-economic consequences. In spite of those facts, we can say that, in the majority of its areas, the Mediterranean Sea has preserved its purity and beauty. It is less polluted than, for example, the Baltic, North or Black Seas. There are, however, numerous "black spots": estuaries of big rivers, zones around large urban agglomerations, and many semi-enclosed areas. In general, north-western regions are more polluted than the south-eastern ones. The present tendencies, however, show the improvements in the former and deterioration in the latter.

A consequence of the pollution of marine environment is a dramatic change of marine biotic systems (e.g., malformed and diseased fish) which in some cases requires the enforcement of restrictions on the use of fish for human consumption. Another phenomenon is the accumulation of toxic substances in sediments of estuaries, deltaic waters, harbour and city ports which enter the food chain. In some cases of semi-enclosed marine bodies, even after the pollution sources have been eliminated, the bottom sediments remain the sources of pollution for longer periods.

Due to the development of tourism and recreational activities many coastal beaches are exposed to overbuilding of hotels and tourist establishments (causing additional pollution), and to exploitation far exceeding their carrying capacity. The often inappropriate behaviour of tourists in environmentally fragile areas, historic settlements and archaeological sites also causes serious damage. Furthermore, it should not be forgotten that the nautical tourism is one of the most propulsive and commercially interesting, but also one of the most aggressive types of tourism activities.

Lagoons, wetlands and marshes are often drained for agricultural purposes, and exposed to felling and elimination of their rare species (mangroves, for example).

As a frequent consequence of human interference in natural coastal processes, there are various forms of coastal erosion, dune destruction and beach modification.

Analysis of these and other pollution processes and irrational use of marine resources and coastal zones led to the awareness that the marine resources and those of coastal areas are specific systems which have to be treated in an entirely new manner.

## INTEGRATED COASTAL ZONE MANAGEMENT

The described situation, as well as recent scientific research brought about a number of new facts. In the first place, it was realized that it is coastal zones and the adjacent sea, and not the regional seas that are most threatened by the uncontrolled development and pollution. It was understood that each eco-system has a limited capacity of assimilating the pollution load, that above that load very intensive processes start (red tide, plankton bloom, eutrophication, mucuous), and that those are extremely fragile systems the recovery of which, after the damage has been done, is a long and expensive process. It has finally become evident that, due to the number, complexity and intensity of relevant interactions occurring in coastal zones, the classical, sectorial approach could not satisfy the requirements of environmental protection and rational use of resources. Since the above problems result mostly from the uncontrolled development processes, the need was universally accepted of integrating the environment policy in the development policy, and applying an integrated approach to the coastal zone management.

The awareness of the specific nature of coastal zones and that very term, "coastal zones", in the context of management appeared in late 1960s in the USA, and especially with the enforcement of the Coastal Zone Management Act in 1972 [2]. It was a result of the awareness of a special nature of coastal zones, the intensity of interactions, and the high value of resources, as well as of the need to manage and protect such areas using methods so far considered unconventional.

A coastal zone can be defined as an organic whole consisting of littoral area, coastal strip and its adjacent sea area. One of the first definitions of the coastal zone was provided by the US Commission of Marine Science, Engineering and Resources, according to which a coastal zone is made by "... the interface of transition that part of the land affected by its proximity to the sea and that part of the ocean (sea) affected by its proximity to the land"[3].

Application of this, or any other definition of a coastal zone meets difficulties due to the fact that many important influences act over large areas and at long distances. Therefore, in the process of coastal zone management a special attention has to be paid to the identification of the coastal zone boundaries, and relevant inland interactions.

## Concept of Sustainable Development

The concept of sustainable development has been formulated in the report of the World Commission on Environment and Development: "*Sustainable development is a development that meets the needs of the process without compromising the ability of future generations to meet their own needs*". A similar definition was given by the OECD: ... *the key to a sustainable development is "... resources use maintaining sustained yields from national resource stocks and meet the needs of future generations*".

The objectives of a sustainable development could be described as follows:

- to achieve a needed rate of development;

- to change (improve) the quality of development;

- to secure basic needs for:
    - employment
    - food self-reliance
    - energy and water
    - health conditions
    - education and social security

- to protect and improve the use of resources (maintain biodiversity, protect endangered species, protect natural and cultural heritage);

- to improve technologies and manage the risks.

The principal elements of sustainable development could be listed as:

- sustainable use: exploitation of renewable resources at a rate equivalent to the regeneration rate

- sparing use for non-renewable resources: recycling and improvement of technologies

- substitution: for both renewable and non-renewable resources - changes in the supply and demand

- compensation: keeping the total amount of natural and man-made resources constant

The principal dimensions of sustainable management of resources could be listed as follows (1):

- economic and ecological integration;

- intergenerational equity: equal and fair distribution of welfare over generations;

- interregional links and trade-offs, openness to cross-boundary flows, external factors and trade-offs;

- multiple use - a number of possible uses of resources - simultaneously, sequentially;

- long-term uncertainty: chaotic behaviour, influence of natural and man-made catastrophes, risks and surprises.

Such a development concept requires a new way of planning and management, as well as a new way of economic thinking: instead of the standard ways of research into the assessment of GNP, it is necessary to include the elements of environmental accounting: values of resources depletion, of environmental damage, environmental rent. The traditional way of benefit/cost calculation has to be replaced by the environmental way: benefits and costs have to be calculated for performing environmental protection activities. It can be said with certainty that environmentally sound economic considerations can avoid or at least diminish the risk of damage resulting from the traditional way of planning. Such new way of thinking requires the application of new measures of environmentally sound management, such as: pollution taxes, tradable emission permits, assurance bonding, and incentive based approach.

## Integrated Coastal Zone Management
Integrated coastal zone management (ICZM) can be defined as a set of activities implemented at political, scientific and

technical, and administrative levels with the objective to formulate, plan, implement, monitor and re-evaluate the process of sustainable development and environmental protection of coastal zones.

The main feature of the ICZM is an integrated approach based on system analysis. It is, therefore, a long-term process with feed-back character.

As a wider and perhaps more flexible definition we could quote that *"... coastal management ... (is) ... any governmental programme established for the purpose of utilizing or conserving a coastal resource or environment ... intended to include all types of government interventions"*[2], where we have to point out, however, the omission of the term "integrated", the differences between the programme and feed-back process, and linking to governmental interventions.

CAMPNET Workshop (Charlestone, USA, July 1989) defined the ICZM as *"... a dynamic process in which a coordinated strategy is implemented for the allocation of environmental, socio-cultural and institutional resources to achieve the conservation and sustainable multiple use of the coastal zone"*[3].

The ICZM is a complex mechanism composed of a number of systems and sub-systems which are in inter-action. Understanding that mechanism implies systematic studies and knowledge, in the first place, of the natural eco-system. Of a particular importance in that context is the <u>institutional arrangement of management elements</u>: the political system, laws, standards, norms, administrative organization and customs. The present institutional arrangement, within which the coastal zone mechanisms act, has to be distinguished from the institutional arrangement of the programme or project which is composed of a system of selected institutional elements needed for its implementation.

The <u>prerequisites</u> of the coastal zone management are the knowledge of the coastal and marine eco-system concerned, the state of environmental pollution, as well as a socio-economic, institutional, political and other interactions influencing that system. It is also necessary to know and understand the influence of the zone on the immediate and faraway areas, as well as influences of transboundary phenomena (pollution impacts, navigation and transport, fisheries, migration of marine and coastal organisms, etc.). However, these requirements imply the existence of relevant management-oriented systems of research, monitoring and survey of pollution. Finally, knowledge of integrated planning and of planning and management tools, applicable in country and coastal zone real conditions increases the efficiency of coastal zone management and its chances for success.

In developing the ICZM system, each country and each coastal zone will set its own global and specific <u>goals and priorities</u>. The structure of the ICZM system will develop in accordance with so defined goals. Although the final decision on the goals and priorities within a coastal zone management system is always site specific, some general goals may be

common for a large number of cases. In this context, the US
Coastal Zone Management Act will be mentioned once again. It
declares for the United States the following four basic
coastal management policies:

- To preserve, protect, develop, and where possible, to
  restore or enhance the resources of the coastal zone;

- To encourage and assist the state to develop and to
  implement CM programmes meeting specified national
  standards;

- To encourage the preparation of "special area management
  plans" to protect nationally significant natural
  resources, to ensure "reasonable coastal-dependent
  economic growth", and to provide "improved protection of
  life and property in hazardous areas and improved
  predictability in governmental decision-making";

- To encourage the participation and the cooperation of
  public, state and local governments, inter-state and
  other regional agencies, and federal agencies in
  achieving the purpose of the CZMA.

According to the author's experience, under the
conditions in the Mediterranean developing countries, an
integrated coastal zone management system should contain the
following:

- a CZM-oriented research of the eco-system, pollution
  monitoring and survey programme;

- a set of measures for immediate and short-term activities
  to protect coastal systems and reduce pollution;

- conditions, measures, and actions for sustainable coastal
  resource uses and development;

- measures to ensure appropriate and harmonized management
  of the most important (sectorial) resources;

- measures to protect and restore areas of exceptional
  natural, historic, cultural and /or other values;

- measures to assess and manage/mitigate natural and other
  hazards;

- a programme of development and improvement of national
  and local capabilities for research, monitoring, planning
  and management of coastal zones;

- an institutionalized system of public participation in
  the process.

The functioning of the ICZM system should, among others, secure:

- development of integrated planning capacities capable of preparing, among others, the preliminary studies and forecasts with a support of modern and applicable tools;

- cooperation (wherever possible) and assistance (if necessary) of relevant international organizations and institutions;

- preparation of integrated plans of development or, if this is not possible, integrated studies on the basis of available data and information;

- necessary legislation (legal acts, standards, etc.);

- financial prerequisites and resources for the plan implementation;

- determination of management policies and procedures, and formulation of legal decision(s) for their enforcement;

- implementation, through management activities, of adopted plans and decisions;

- supervision and control over the plan implementation, monitoring of results, and re-evaluation of goals, policies and plans.

The basic requirement of such an institutional arrangement of a project or a programme is that the following has to be secured:

a) vertical integration of management levels (national, sub-national, local);

b) horizontal integration (competent authorities and institutions); and

c) involvement of not other than the directly relevant factors.

Placing the programme implementation at the optimum level is also of a great importance, that level being, as a rule, the lowest level securing the programme implementation.

Due to their specific features and high value, the coastal resources, coastal zones and Exclusive Economic Zone should, as a rule, be under control of public authorities, primarily the government, while for the immediate sea shore and shorelands it is necessary to secure extensive governmental or delegated to lover levels public control.

## Integrated Planning, a Major Tool of ICZM

The new understanding of the nature of coastal zones, of sustainable development which secures a rational development with the protection of resources and environment, and of the need to apply ICZM led to a new approach to planning, especially of coastal zones. Conventional, sectorial, planning methods resulted not only inadequate, but also a cause of uncontrolled development and environmental pollution. Therefore, especially in developed countries, a concept of integrated planning was introduced and applied, where the term "integrated" refers to the integration of technological, social, cultural, ecological and institutional elements into a planning process which, like the ICZM, has a permanent feed-back character.

Ever since the beginning of its activity, the Priority Actions Programme of the Mediterranean Action Plan has been implementing, among others, the priority action "Integrated Planning and Management of Coastal Zones". The main objective of that action is to develop, on the basis of the experience of the Mediterranean countries, especially the developed ones, a methodological approach to the planning of coastal zones, primarily under the conditions of the Mediterranean developing countries. Several years of work of a number of renowned Mediterranean experts resulted in the formulation, and finally adoption, of "A Common Methodological Framework for Integrated Planning and Management in Mediterranean Coastal Areas"(4). According to that document, *integrated planning is a dynamic process of achieving goals and objectives for environmentally sustainable development, within the limits of physical, social and economic conditions and within the constraints of legal, financial and administrative systems and institutions. Being process-oriented, integrated planning does not have as its objective the preparation of an ultimate product-plan, as an ideal state which is hoped to be achieved by a certain time in future. Rather, it is an adaptive process of resource management, capable of responding to expected or unforeseen changes and events. This process includes analysis and forecasting, plan making and evaluation, monitoring and feedback, all of which should be oriented towards achieving clearly defined goals and objectives through practical and effective means of implementation.* A schematic presentation of the process of integrated planning is given in Fig. 2.

That methodological framework served as the basis for the development of PAP pilot projects and, later, MAP Coastal Area Management Programmes (CAMPs), and its applicability and flexibility were verified through the application in a number of Mediterranean countries under various general and specific conditions of planning and management.

It can generally be said that the application of integrated planning is particularly necessary for coastal areas, where dynamic on- and off-shore natural processes are constantly changing physical conditions, where intense conflicts arise between economic activities competing for land space, shoreline or sea space, and where development at one

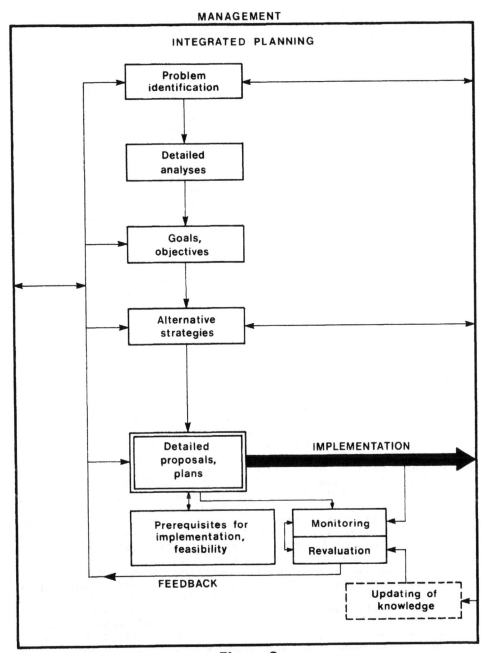

**Figure 2**
**METHODOLOGICAL FRAMEWORK FOR THE PROCESS
OF INTEGRATED PLANNING OF COASTAL ZONES
-PHASES OF THE PROCESS-**

124

location may generate significant changes in conditions elsewhere. It implies a multidisciplinary approach, environmental evaluation and socio-economic and financial appraisal.

It is important to point out the fact that the concept of ICZM is not in contradiction with the directives of the EC or the regulations of the developed Mediterranean countries.

## Environment-Development Scenarios - the Blue Plan

The multidisciplinary nature of development-environment interactions, and complex requirements imposed by the ICZM and IPCZ called for the development of numerous specific tools. In the Mediterranean conditions, those tools, created mainly in the developed countries, had to be tested and modified in order to be applicable in the Mediterranean conditions. With that in mind, the Blue Plan RAC applied the technique of development-environment scenarios for the preparation of the prospective study "The Blue Plan: Futures of the Mediterranean Basin"(5).

The prospective studies explore into possible futures, using different scenario methods. The objectives of those studies are: to analyze the interaction mechanisms of the area in consideration; to analyze the likely evolution of environment-development interactions; and to illustrate possible consequences and breaking points.

An environment-development scenario could be understood as a link between the present and the future through a pathway built in stages of 5-10 year periods.

Such studies provide the decision-makers, planners, managers and the general public with an opportunity to understand possible development alternatives with their prerequisites and consequences. However, it has to be pointed out that those are neither predictions nor forecasts, and under no conditions can be identified with plans.

The results of the MAP Blue Plan are presented in the final report and summarized in a separate publication, while the information and data relevant to the region are systematized in the Blue Plan Data Base. In addition, 18 specialized booklets are being prepared for various economic sectors or geo-political environments. After several years of studies in which a number of experts from all Mediterranean countries took part, three different scenarios were developed and analyzed: rapid growth trend, weak growth trend, and alternative, balanced economic growth trend.

The analysis of the 3 scenarios suggested that the development of the region at a high growth rate would require massive financial resources which would not leave much for the necessary protection of the environment and ultimately lead to its considerable deterioration. This would mean tremendous losses and much larger investments later to improve the state of the environment, if the improvement would be possible at all.

The weak growth trend scenario rendered unstable due both to the increasing deterioration of the socio-economic situation which would struck some Mediterranean countries and an increasing degradation of the environment and the resources.

Both trend scenarios lead to an unsatisfactory future and should not make the basis of a future development.

The main goal of the alternative trend scenario is the achievement of a balanced development with environmental considerations as part of national development strategies. One option of this scenario ("Mediterranean") presumes the existence of a strong and well organized Mediterranean cooperation with the CEC as a promoter of a sustainable development, while the other one is based on a somewhat weaker influence of the CEC, presuming intensive linking at sub-regional levels (Maghreb, Arabian east, north-Mediterranean regions). The first option envisages a somewhat faster growth than the other. Both options presume an integrated approach, i.e. acting simultaneously on all resources, including the human potentials.

With regard to the importance of the results of the Blue Plan for the analysis of development options of the Mediterranean, the main indicators are given in Fig. 3.

## Other Tools and Techniques of Integrated Planning and Management of Coastal Zones

Within the priority action on integrated planning and management of coastal zone, PAP adapted and tested through pilot applications a number of other tools and techniques, of which the following should be mentioned:

- application of GIS at PC level using ARC/INFO software;

- a practical approach to Environmental Impact Assessment[6];

- carrying capacity assessment for tourism establishments[7];

- application of multicriterial analysis for optimum site selection for development projects[8];

- hazard assessment and risk management of industrial, energy, transport and other activities.

The MAP Coordinating Unit and UNEP-OCA/PAC have recently developed and tested a method of practical cost/benefit analysis of the application of measures for the protection against pollution from land-based sources.

Actions related to the application of the Protocol for the Protection of the Mediterranean from Accidental Pollution are implemented by the MAP Marine Pollution Emergency Response Centre (REMPEC) in Malta. Within that programme, training programmes were developed, a Regional Information Centre was established, and a number of documents were prepared regarding the operational arrangements.

| | |
|---|---|
| Population | 360 million in 1988 |
| | 520-570 million in 2025 |
| | |
| Urban water consumption | |
| Increase in 2025 as compared to 1985: | 40-60% (N) |
| | 400-500% (S,E) |
| | |
| Soil | 300 milt/year loss of fertile sediment |
| | 1/3 fertile land affected by erosion |
| | |
| Forests | to be reduced (up to 40%) and degraded |
| | |
| Coastline | up to 2025 additional loss of 4000 km |
| | due to urbanisation |
| | |
| Industry | strong development in the south: |
| | pollution |
| | |
| Total energy consumption | 0,6 billion tons oil equivalent in 1988 |
| | 1-1,5 billion tons oil equivalent in 2025 |
| | |
| New 150-200 thermal power plants | |
| or equivalent NPP | |
| | |
| Number of tourists | 108 million in 1984 |
| | 380-760 million in 2025 |

Consequences: pollution, overurbanization, water scarcity, soil degradation, threat to genetic, cultural and natural heritage

Remedies: international cooperation, orientation to sustainable development through integrated planning and management of resources

### Figure 3
### FUTURES OF THE MEDITERRANEAN BASIN
### MAIN FINDINGS OF THE BLUE PLAN

The Regional Activity Centre for the Specially Protected Areas in Tunis prepared the criteria for the identification and protection of areas of special natural, historic and archaeological importance, and an inventory of the hitherto included areas on the Mediterranean.

All of the above tools and techniques were applied in a number of Mediterranean countries within pilot studies or applications, and especially within MAP CAMPs.

## Climate Change and Coastal Zone Management

Assessment and mitigation of impacts of expected climate changes is a very important new element of coastal zone management. Research performed in the last 10 years confirmed as highly probable that in the next century the World will have to face significant climate changes that will have a strong and unfortunately predominantly negative impact on the Mediterranean coastal zones. Of the numerous scientific studies, documents and meetings which dealt with the problems of the expected climate changes and their impacts, we shall mention the activities of the Intergovernmental Panel on Climate Change established by UNEP and IMO, and especially the results of the Second World Climate Conference held in Perth in 1990.

As a working hypothesis for the assessment of the impacts of climate change, at least for the northern and eastern parts of the Mediterranean, the Perth conference took the following elements, taking 2050 as the planning time horizon:

- sea level rise 24-52 cms;

- mean temperature elevation 1.5-3°C;

- statement that, concerning potential changes to the climate of Southern Europe (35°-50°N; 10°W - 45°E): "warming would be 2°C in winter and would vary from 2°-3°C in summer; summer precipitation would decrease by 5-15% and summer soil moisture by 15-25%"

UNEP and MAP had organized the preparation of a large number of studies of impacts of the expected climate changes which were presented and evaluated at 2 meetings of the Task Teams on Implications of Climate Change in the Mediterranean, Caribbean, South-East Pacific, East Asian and South-East Asian Seas (Split, 1988, and Perth, 1990). At the same time, PAP prepared a methodological framework for integrating the assessment and mitigation measures of climate change impacts into the integrated coastal zone planning and management process in the Mediterranean conditions. The numerous studies prepared by MAP ( on the impacts on the Venice Lagoon, Ebro Delta, Nile Delta, Amrakikos Gulf, Rhône Delta, and other areas) confirmed the significance and importance of bearing those changes in mind already now within the process of ICZM, although a more significant impacts of those changes will only be felt in 30-40 years. This particularly refers to the planning of major projects, infrastructure network and development of towns in flat coastal areas.

Departing from that experience, MAP included in the CAMPs the element of assessment and mitigation of climate change impacts.

Of the numerous studies and documents on climate change impacts particularly interesting for the Mediterranean, we shall mention "Multicriterional Analysis for Environmentally Sound Siting of Development Projects Applicable in the Mediterranean Region" and "Report on the Joint Meeting of the Task Team on Implications of Climatic Changes in the Mediterranean and the Co-ordination of Task Teams for the Caribbean, South-East Pacific, East-Asian and South-East Asian Seas, (UNEP(OCA)/WG.2/25)[8,9].

## Integrated Coastal Zone Management Programmes (ICZMP)

An Integrated Coastal Zone Management Programme can be defined as a comprehensive set of activities aimed at achieving determined objectives within the process of integrated coastal zone management. In the context of this document it is understood that a CZMP is based on the principles of sustainable development and that it uses the integrated planning and other applicable tools and techniques of ICZM. At that it is necessary to clearly distinguish the term "programme" from the term "project" since a CZM programme can include a number of different projects.

Essential elements of an ICZMP are: boundaries, strategy and objectives, institutional arrangements, contents and duration, funds for the implementation, and finally, programme evaluation.

Programme boundaries. The area of an ICZMP includes a coastal strip (shore line), an inland zone and the adjacent sea. Inland boundaries have to be defined with regard to the intensity of interactions between the immediate coastal strip and the adjacent sea, but it is necessary to take into consideration the geographic, climatic, morphological, administrative and other elements. It is practical if the zone coincides with one or more administrative and statistical units. In all cases, it should be expected that some interactions will be felt outside the set boundaries. Such interactions have to be identified, evaluated and taken into consideration when formulating and implementing the programme. With regard to the development character of the ICZMP, inland boundaries have to include the relevant development centres. Seaward boundaries may vary in dependence of the adopted approach or specificities of the area, that is several miles from coastline, or up to continental ridge, or to the limits of the exclusive economic zone, while some states have regulations for those matters.

Strategies of ICZM. Each ICZMP requires the definition of an adequate strategy and specific programme goals. Many basic elements of the strategy and some of the objectives are common to all coastal zones while others may be specific to one or several coastal states, regions or areas sharing similar conditions.

Two documents that could be considered as defining strategies for all Mediterranean countries are the Genoa Declaration and the Charter of Nicosia.

The Genoa Declaration was adopted at the 5th Ordinary Meeting of the Contracting Parties held in Genoa in 1985. It presents the basic strategy of MAP for the period 1985-1999. Among others, the Declaration defines 10 priority targets to be achieved until 1995 relative to: the establishment of port reception facilities for ballast waters, construction of sewage treatment plants in all towns with more than 100,000 inhabitants, and outfalls and/or treatment plants for the towns with 10,000 - 100,000 inhabitants; application of the environmental impact assessment; improvement of the safety of navigation; protection of the endangered marine species; substantial reduction of industrial pollution and solid waste; protection of historic sites; measures to prevent forest fires, soil loss and desertification; and substantial reduction of air pollution.

Five years later, in a meeting convened jointly by the CEC and UNEP-MAP in Nicosia, Cyprus, the representatives of all Mediterranean governments and the CEC adopted the Charter of Nicosia. Departing from the deliberations of the Genoa Declaration, this Charter defines a number of common commitments, such as:

- to achieve, as a fundamental objective for the year 2025, an environment in the Mediterranean compatible with sustainable development;

- to elaborate and adopt environmental management strategies as an integral part of the socio-economic development of the countries;

- to implement EIA of development projects;

- to implement economic and fiscal incentives and disincentives, and administrative measures to improve the integrated management of the environment; and

- to carry out the priority actions on: integrated management of coastal zones; setting up installations for the treatment of municipal wastewater (25 of which will be built by 1995 with the financial assistance of the CEC); providing by 1993 at least 20 Mediterranean ports with equipment for ballast and bilge waters treatment; increase of exchange of experience in setting up of standards and appropriate environment management institutions, and protection of semi-enclosed seas and their coastlines in the Mediterranean basin; organization of training courses for managers and technicians in the priority areas; promotion of national campaigns for energy and water savings and management of non-renewable resources.

The Genoa Declaration and the Charter of Nicosia can be considered fundamental documents which establish the strategy of the protection of the Mediterranean adopted at the level of the Mediterranean governments and the CEC.

Objectives of ICZM. When defining the objectives it is necessary to distinguish the long-term from the immediate ones. At that, it has to be started from the desired and necessary to the possible. One of long-term objectives of each ICZMP in the Mediterranean has to be strengthening the capabilities of the national and local institutions for the implementation of the process of integrated planning and management of coastal zones.

Institutional arrangement. Structure of authorities and management of each country is a result of historic, political, socio-economic and numerous other factors which are almost always country-specific. In the majority of Mediterranean countries, the authorities directly or indirectly involved in the ICZM are organized in 15-25 organizationally separate sectors each subordinated to at least one governmental agency and/or ministry. At that, each sector has several functions, from data collection to regulation. Mathematically, that leads to 150-200 points of potential involvement in ICZM. It is, therefore, necessary, in the preparation of an ICZMP, to perform a study of the system and general efficiency of the institutional arrangement, and of the participation of the public and non-governmental organizations participation in it. Such a study is not only a prerequisite of the formulation and efficient implementation of the programme, but it can, as an output of the programme, instigate the authorities and decision-makers to look for and introduce in a future institutional arrangement improvements in the coordination of authority in horizontal and vertical linking of their functions.

Whenever possible, the institutional arrangement of a programme should include relevant international institutions and/or UN agencies that could secure training, assistance in the formulation and implementation of the programme, and transfer of international knowledge and experience. When selecting the institutional arrangement for an ICZMP it is necessary to secure the participation of or harmonization with the necessary number of authorities and institutions. The programme management mechanism has to be institutionalized at the lowest level which can guarantee an efficient and rational implementation. Among applied solutions there are: inter-ministerial bodies, multisectorial and multilevel commissions or authorities, or, on the other hand, entirely new, *ad hoc* institutions given executive and control authority. Arrangements of lasting character could have precedence, under the condition that an adequate and clear distribution of responsibility and authority is achieved.

The usual organizational and process problems to be taken into consideration when preparing an ICZMP are:

-       lack of information and database, especially a lack of knowledge on the ecosystem(s) and its assimilative capacity;

- insufficient institutional and human professional and scientific capacities, especially insufficient knowledge of the principles of sustainable development, ICZM and integrated planning, and omission of modern tools and techniques;

- lack of coordination among authorized organizations;

- deficiency of legal regulations and its poor implementation;

- decisions made on sectorial considerations, excluding the environmental ones;

- lack of clearly stated strategies, policies and goals; predominance of short-term objectives;

- lack of funds or unharmonized cash flows;

- limited public participation or its total absence;

- in some developing countries, unfortunately very often, public properties are improperly managed.

Duration of ICZMP. Bearing in mind the complexity of the coastal zone management mechanisms, the still insufficiently affirmed principles of sustainable development, and the length of the process of protection against pollution and recovery of polluted areas, an ICZMP should have, as a rule, at least a medium-term character. Most of the programmes implemented by USAID (US Agency for International Development) last 8-10 years. With regard to their specific nature, the MAP programmes last 4-5 years, including the preparatory phase. Because of the long duration of ICZMPs, it is advisable to reach a general consensus in advance, in order to avoid possible political implications later, in occasions of elections or change of local and/or national authorities. It is also important that ICZMPs have intermediate outputs sufficiently obvious for the general public and useful to decision-makers. Those have to include outputs which secure a certain improvement of socio-economic conditions and quality of living. That should maintain the interest and support of the authorities and the general public to the programme and its objectives.

Financing of the programme. Many programmes failed or could not be competed due to unrealistic budgets, insufficient funds and their untimely availability. It is, therefore, necessary to secure realistic and stable sources of financing for the implementation of the programmes, where in general, the principle is applied of cost sharing between national and local authorities, and, if possible, international donors or creditors.

<u>Structure and contents of a programme</u> have to be harmonized with its strategy and objectives, the secured funds, and, which is particularly important, with the realistic assessment of the possibilities of its implementation (institutional, scientific and professional capacity at the national and local levels, data availability, and level of environmental knowledge). In order to secure the efficiency of the programme, it is recommendable to first start a preparatory phase to create the necessary conditions and to collect and study the available data and information, as well as to establish the institutional mechanism and secure its capacity necessary for the implementation of the programme, to define the strategy and objectives, and to secure the financial support. The main part of the programme has to be divided in phases, and the outputs of each phase have to be defined. Finally, evaluation of the programme has to be secured in the implementation phase, and after its completion. With regard to the situation in the majority of Mediterranean developing countries, already the implementation of the prerequisites mentioned here for the preparatory phase can be understood and accepted as a phase of CZMP.

<u>Programme evaluation</u>. In the implementation phase and after the completion of a programme, it is necessary to secure its evaluation. In the implementation phase, that evaluation can be performed at a yearly basis with the objective of analyzing the efficiency of the programme and identifying possible necessary changes. At the same time, the implementation manner and outputs of the programme also have to be evaluated. After the completion of one phase or the entire programme it is necessary to examine, among others, whether the objectives were realistic and clear, whether the implementation was rational and efficient, whether the budget was realistic, and whether the evaluation of results and outputs was good. In the USA, the evaluation of ICZMPs is a standard practice, performed by the Office of Coastal Resources Management. UN agencies also perform the obligatory evaluation during and after the completion of the programmes.

## MAP COASTAL AREA MANAGEMENT PROGRAMMES (MAP CAMPs)

Ever since the beginning of its activities in the period 1976-78, MAP has been aware of the need to apply the process of integrated management of coastal zones and integrated planning in the process of environmental protection of the Mediterranean. Such orientation was gradually applied, first within PAP, through the priority action "Integrated Planning and Management of Mediterranean Coastal Zones", and after the necessary conditions were created, through the PAP pilot project implemented in the period 1987-89. On the basis of a proposal by the Executive Director of UNEP, the 6th Ordinary Conference of the Contracting Parties (Athens, 1989) approved, among others, the implementation of 4 MAP integrated coastal

area management programmes in the areas which had already
passed the preparatory phase within the PAP pilot projects
(the Syrian coast, the Izmir Bay, the Kastela Bay, and the
island of Rhodes). At the same time, the Conference approved
the concept, programme and manner of implementation of MAP
CAMPs.

## MAP Approach to Coastal Zone Management

MAP CAMPs were defined as specific programmes implemented
jointly by MAP (with the participation of all its components)
and national and local authorities and institutions.

The following criteria were determined for site
selection:

- it has to be a set of sites of various types within the
  typology of the Mediterranean coastal zones;

- the selected zone has to be either threatened or already
  significantly affected by and polluted due to
  uncontrolled development;

- there have to exist sufficient capacities of national and
  local institutions and authorities that can carry the
  implementation of the joint programme;

- those authorities have to be interested in the
  implementation of the programme, or in some actions
  at that site already in course or in preparation;

- the jointly achieved results have to be available for
  use, as pilot experience, in other areas.

For the definition of boundaries, criteria were accepted
as mentioned in the chapter 2 of this document. In the
selection of the first 4 sites, there was a favourable
circumstance of administrative boundaries satisfying all other
criteria.

When defining specific objectives and strategies of the
programmes, the departure points were the Genoa Declaration
and the Charter of Nicosia (see point 3.7 of this document).

The long-time cooperation of MAP and the governments and
institutions of the Mediterranean countries in which the
selected areas are located, and the experience gained by PAP
pilot projects excluded the need of performing, in the
preparatory phase, detailed analyses of general conditions in
those countries and particularly the institutional
arrangements. Detailed analyses of specific conditions were
performed within the PAP pilot projects.

When defining the conceptual framework and approach, it
was necessary to take into consideration the general MAP
principles of action and some regulations of the Convention.
The fact that UNEP and MAP are not funding agencies resulted
in an approach by which the MAP contribution to the
implementation of CAMPs consists of the outputs, experience

**MAP-Coastal Area Management Programmes (CAMPs)**

**Definition:**
A form of advanced collaboration with national and local authorities and institutions based on principles of sustainable development and integrated coastal zone management

**Main objectives:**       to introduce or develop the process of integrated planning and management of Mediterranean coastal zones, to contribute to a sustainable development and environment protection

**"Bottom-up" approach:**
- methodological framework
- pilot areas
- wider application of results and experiences obtained in pilot areas

**CAMP phases:**

**1. Preparatory:**
- data collection
- upgrading of capacities
- environmental knowledge (assimilative capacity, identification of problems and climatic impacts)
- programme formulation

**2. Implementation:**
- data base
- training
- coastal zone scenarios (development, climatic changes)
- integrated planning studies (resource evaluation, impact assessment, development outlook, immediate and long-term mitigation measures
- programme of an integrated plan

**3. Follow up:**
- preparation of an integrated plan
- implementation
- monitoring
- re-evaluation

**Figure 4**
**METHODOLOGICAL FRAMEWORK FOR MAP COASTAL AREA MANAGEMENT PROGRAMMES**

and network of its experts and institutions, performing the following functions: completion of environmental knowledge; promotion of the principles of sustainable development and ICZM; application of modern tools and techniques and regulations of the Barcelona Convention; training; exchange and application of updated international knowledge and experience. MAP also provides the assistance of international, primarily Mediterranean experts who cooperate as consultants with national and local institutions and experts. A scheme of the methodological framework of MAP CAMPs is given in Fig. 4.

The focus of the programme implementation is on the work of local institutions and experts. This approach was selected since the analysis of a large number of international programmes implemented in Mediterranean developing countries by highly competent international expert teams failed to have an efficient follow-up due to the fact that the local institutions and experts had not been sufficiently involved in those programmes and their up-grading had not reached a level which could guarantee a successful use of the programme results.

With regard to the fact that the Contracting Parties of the Barcelona Convention evaluate the completed and approve the future programmes and budgets on a biennial basis, the workplans of MAP CAMPs also had to be defined in two-year periods. Thus the structure of MAP CAMPs is defined in 3 phases:

- preparatory phase, carried out mostly by PAP and MEDPOL, which includes collection of data, acquisition of knowledge on local and national capabilities and institutional arrangement, networking, creation of conditions for national and local co-financing, and, if possible, for external financial support, and, finally, formulation of the programmes and of the agreement to be signed; this phase lasts for one biennium;

- implementation phase in which most part of the programme is implemented and which also lasts for one biennium;

- final phase in which the results are analyzed, activities which could not be performed in the implementation phase are completed, the final report is prepared and the results presented at the local, national and international levels.

This shows that the duration of a programme is approximately 5 years, i.e. only the lower limit is reached of duration recommended for such programmes. The Fig. 5 presents the MAP approach to CAMPs.

Institutional arrangement for the implementation of MAP CAMPs comprises all MAP components coordinated by the Coordinating Unit in Athens, the authorized national institution acting as the National Focal Point for MAP (usually a ministry or national agency for environment), and

CONTENT OF  MAP - C.A.M.P.s

1  IMPLEMENTATION OF LEGAL INSTRUMENTS: LBS Protocol (monitoring, survey of pollution), MARPOL, Emergency and Dumping Protocol

2  RESOURCE EVALUATION, PROTECTION AND MANAGEMENT: water, soil, forests, coastline, marine ecosystems, protected areas

3  ACTIVITIES - evaluation and trends

4  NATURAL HAZARDS AND PHENOMENA: seismic risk, impact of climatic changes

5  PLANNING AND MANAGEMENT TOOLS: database, GIS, EIA, Carrying Capacity Assessment for tourism activities

6  INTEGRATED PLANNING AND MANAGEMENT:

Integrated planning studies

Resources protection and management plans

Development-environment scenarios

INTEGRATED PLANS

↓

IMPLEMENTATION, MANAGEMENT

MONITORING

FEEDBACK

**Figure 5**
**MAP APPROACH TO COASTAL MANAGEMENT**

the local authorities in charge of programme implementation.
For each activity or project within a MAP CAMP, a separate
institutional arrangement is defined including other relevant
ministries, authorities and institutions.

The **legal bases** of the implementation of a MAP CAMP are:
(a) deliberations of the Conferences of the Contracting
Parties; (b) the contract signed by the the MAP Coordinator
and the authorized person (minister, director, president) of
the national institution acting as the National Focal Point
for MAP. Such contracts were made for each of the 4 on-going
MAP CAMPs.

**Cooperation with the World Bank and European Investment
Bank**. After the Agreement on Joint Activities and Cooperation
in the Implementation of Coastal Zone Management Programmes
was signed by UNEP and those banks in 1988, MAP established a
direct cooperation with them within the implementation of MAP
CAMPs. On the basis of that cooperation, those banks agreed
that, apart from their own programme of help in the
Mediterranean (EPM-METAP), they would co-finance the on-going
MAP CAMPs in the amount of up to US\$ 300,000 per programme.

**Financial arrangements**. The contracts for each CAMP
define the financial liabilities of each participant. Although
there are some variations in cost-sharing structure, it could
be said that in general 50% of the expenses, in cash and kind,
are covered by the national and local institutions and
authorities, and 50% by MAP and international donors.

**Outputs and benefits** of each CAMP are defined with regard
to the programme. For each individual activity separate
outputs are defined, while for the overall programme the
general outputs are the fulfillment of the general outputs and
the final report.

**Programme evaluation** is envisaged with the following
dynamics: yearly (at the meetings of the MAP techno-economic
committee), biannually (at the ordinary conferences of the
Contracting Parties), and upon the completion of the
programme.

## Description of Selected Areas for the Implementation of MAP CAMPs and their Programmes

With regard to the purpose of this document, only short
descriptions will be given of the selected areas and their
programmes, while only the programme of the Syrian coast will
be described in somewhat greater detail.

MAP CAMP **"The Bay of Izmir"**. The project covers the area
of the Metropolitan Municipality of Izmir located around the
Bay of Izmir in the Aegean region of the Western Coast of
Turkey. The population amounts to 1.5 million with a tendency
of constant growth (4.4% annually). The city is a strong
industrial centre, a commercial and navy harbour. Most of the
industrial and all urban wastewater and run-off are discharged
into the bay without any treatment. The bay is heavily
polluted and the Inner Bay faces a progressive process of
eutrophication. Tourist activity is intensive in the

surrounding area. National and local authorities have been making great efforts on improving the situation. A project of urban wastewater collection, treatment and disposal is in implementation, co-financed by the World Bank, scheduled to be completed in 1995.

The long-term objectives of the programme are the following:

-   propose a development concept of the area of Izmir harmonized with the receptive capacity of the environment;

-   create conditions for the establishment of the system of integrated planning and management of resources in the area of Izmir by:

    a)  establishing a monitoring programme of the environment on a permanent basis;

    b)  setting up a data base of all necessary environment and development indicators;

    c)  providing training of local experts on various aspects of the programme.

The objectives of this programme are presented in the Fig. 6. The immediate objective of the programme is to give, within individual actions, solutions of environmental problems of the most urgent nature which could be implemented immediately. In the elaboration of those solutions, a particular attention will be paid to the strategic objectives of the programme.

The main expected outputs of the programme are:

-   proposals for immediate actions;

-   technical and economic measures for addressing existing environmental problems;

-   integrated management plans;

-   studies and reports on specific subjects;

-   training of local and national experts;

-   demonstration projects;

-   monitoring programme;

-   data base for various development and environment aspects;

-   software to be used in solving some specific problems.

MAP Coastal Area Management Programme

## THE BAY OF IZMIR

Objectives: – contribute to the achievement of a sustainable
development harmonized with the receptive
capacity of the environment

– complement environmental knowledge

– create conditions for the establishment of
integrated planning and management process

– train local experts and update institutions on
relevant topics

– mitigate the present pollution impacts and
future impacts of climatic changes

**Figure 6**
**MAP CAMP: OBJECTIVES**

The activities to be implemented by the programme are the following: implementation of the Land Based Sources Protocol, implementation of the emergency protocol and MARPOL Convention, extended monitoring programme of the Izmir Bay, study of environmental capacity of the Inner Bay, study on implications of expected climate changes, training programme on GIS on pcARC/INFO, EIA of submarine outfall, development-environment scenario, integrated planning study of the area, study on the protection of the Tuzla migratory birds nesting area (see Fig. 7).

At the moment of writing this paper, most of the above activities are in the phase of implementation and expected to be completed by the end of 1992, while 3 of them have been delayed due to a delay in providing a part of the financial support for their implementation.

MAP CAMP "The Kastela Bay". This project has been formulated as a contribution to the already on-going national project "Rational Management of Natural Resources of the Kastela Bay" developed and implemented by the University of Split, Croatian Academy of Arts and Sciences, with support by local, national and ECE funds. The project covers the areas of the municipalities of Split, Solin, Kastela and Trogir situated at the Adriatic coast of Croatia, with surface area of approximately 1500 km$^2$ and 350,000 inhabitants. The entire coastal area is urbanized, with numerous industries and well developed tourism on the outskirts. Due to urban growth and development of industry, as well as discharge of only primarily treated wastewaters, the bay is very polluted with eutrophication in its easternmost part. A project of urban sewage collection, treatment and disposal is in the beginning of implementation, but the funds secured so far will suffice only for a part of the sewerage system.

The long-term objectives of the MAP CAMP are the following:

-   to contribute, in as concrete and direct way as possible, to the implementation of the infrastructure programme for the Split agglomeration around the Kastela Bay ("Split-Solin-Kastela Integrated Ecological Project");

-   to contribute to the creation of systems, institutional and local capabilities for establishing a continuing process of integrated planning and management of this coastal area which would enable a harmonious and sound use and development of natural resources, reduction of pollution from the existing sources, and a permanent protection of the bay (in harmony with the "Rational Management of Natural Resources of the Kastela Bay".

The first objective of activities within the framework of MAP implies the collection of a number of important and to date unavailable information as well as the completion of scientific and technical knowledge necessary for the

141

---

### Programme:

1. Survey of land based sources of pollution

2. Measures for prevention of marine pollution by ships, contingency plan

3. Survey and monitoring of pollution in the Bay

4. Study of the Assimilative Capacity of the Bay

5. Study of the Recovery of the Inner Bay

6. Study of Impact of Expected Climatic Changes

7. Study of the protection of Tuzla Area

8. Training on GIS pcARC/INFO

9. EIA of the discharge outfall

10. Development/environment scenario

11. Integrated Planning Study

---

**Figure 7**
**MAP CAMP "THE BAY OF IZMIR"**
**LIST OF ACTIVITIES**

effectuation of the infrastructure programme.

The achievement of the second objective requires the development of modern tools and techniques applicable under local conditions in the process of planning and management, as well as creation of measures and plans of response in cases of accidental pollution.

The main expected outputs of the project are:

- collection of missing ecological and other data and information, and establishment of a permanent monitoring programme;

- studies and reports on specific ecological problems with measures proposed for their solution;

- programme of response to accidental situations;

- application of modern tools and techniques of coastal resources planning and management (geographic information system; assessment of risk from industries, energy and other activities, and risk management; development-environment scenarios; software to be used in solving some specific problems);

- Environmental Impact Assessment of a submarine outfall;

- training of local experts;

- proposals for international financial support for future activities and projects;

- proposal of the follow-up activities.

The programme implementation enjoys a financial support by the World Bank.

Activities implemented within the MAP CAMP are: survey of land based sources of pollution; assessment of risk from pollution by oil and other harmful substances and preparation of a contingency plan; additional pollution survey and monitoring programme; study on the implications of expected climate changes; training on the application of GIS on pcARC/INFO within the planning and management process; hazard assessment and management of risk from energy, industries, transport and other activities; development-environment scenarios; EIA of submarine outfall; study on the optimum treatment level for municipal wastewaters; evaluation of Pantan, an area of specific natural and historic value, study of water resources of the western part of the programme area, study of water supply of the islands of Drvenik Veli and Drvenik Mali.

At the moment of writing this paper, some of the activities have already been completed, and some are in the phase of implementation and should be completed by mid 1992.

MAP CAMP "The Island of Rhodes". The programme covers the Greek island of Rhodes situated at the south-eastern corner of the Aegean Archipelago, at a short distance from Asia Minor. The island covers an area of 1400 km$^2$ with about 90,000 inhabitants. International tourism (20-25% of the total number of international tourists in Greece) is the principal activity in the island. Besides the historic town of Rhodes, in the island there are many other historic settlements and archaeological sites. A strong development of tourism defines the island economy as mono-cultural, and represents a threat through the pollution of the environment, overbuilding, over-exploitation and pollution of freshwater aquifers. The most prominent problems regard the discharge of solid and liquid wastes, protection of historic heritage, and water supply and water resources management.

The long-term objectives of the programme are the following:

- propose a development concept of the island harmonized with the receptive capacity of the environment;

- create conditions for the establishment of the system of integrated planning and management of the island resources by:

    a) establishing a monitoring programme of the environment on a permanent basis;

    b) setting up a data base of all necessary environment and development indicators;

    c) providing training of local experts on various aspects of the programme.

The immediate objective of the programme is to give, within individual actions, solutions of environmental problems of the most urgent nature which could be implemented immediately. In the elaboration of those solutions, a particular attention will be paid to the strategic objectives of the programme.

The main expected outputs of the programme are:

- proposals for immediate actions;

- technical and economic measures for addressing existing environmental problems;

- integrated management plans;

- studies and reports on specific subjects;

- training of local and national experts;

- demonstration projects;

- monitoring programme;

- data base on various development and environment aspects;

- software to be used in solving some specific problems.

The activities to be implemented within the programme are the following: implementation of the land based sources and the dumping protocols; study on liquid waste management, implementation of the emergency protocol; preparation of the General Water Resources Master Plan; study of implications of the expected climate changes; programme of environmentally sound energy planning; programme of protection of historic settlements; training programme on and application of GIS on pcARC/INFO; environmental impact assessment of a sewage treatment and disposal project; development-environment scenario; training programme on integrated planning; integrated planning study for the island; study on the protection of the Butterfly Valley. The programme will be supported by the European Investment Bank.

At the moment of writing this paper only a part of the activities are in implementation, and the programme is expected to be completed by the end of 1993.

MAP CAMP "The Syrian Coast". The programme covers the entire watershed of the Syrian coastal area, with the coastline length of 183 km and population of 1.3 million inhabitants. It is a zone very rich with natural resources and with a very significant development potential. There are several large towns (Lattaquia, Tartous, Banyas) where industry is concentrated. The entire area has sufficient water, the soil is fertile and traditional agriculture had a strong development. Although there are numerous places suitable for tourism, its development has only recently become subject to the interest and plans of the Government. Urban wastewaters are directly discharged into the sea without any treatment. An extensive use of fertilizers causes pollution of the aquifer.

A specific element of the preparation of this programme was in the fact that within the PAP pilot project, among others, a comprehensive planning study of the entire area was prepared.

On a global scale, the development of the region has been to a large extent a response to the basic national needs, In that, the available natural resources (favourable geographical position, mild climate, land and water availability, etc.) have been the major determinants of the economic profile of the region. The existence of these resources has enabled the construction of large transport and industry infrastructure (modern harbours, railroads, highways, oil refinery, terminals and pipelines, cement factory). However, meeting the basic national needs has not been followed by the development of spin-off industries. A somewhat belated regard for agriculture and tourism as major resources was indeed a problem which, paradoxically, has proved to be an asset opening up vast possibilities for the future economic development of the region.

A considerable success achieved in the domain of social development (health, sanitary and educational upgrading) boosted up the population growth so that in the last two decades the coastal population nearly doubled.

The notable development of basic economic activities and infrastructure in the coastal region affected its physical environment having the usual after effects, such as population pressure on the narrow coastal strip and major cities. Concentration of economic activities and population in some parts of the region pushed the physically limited and ecologically fragile resources close to the edge of degradation (pollution of fresh water resources, sea and beaches; soil degradation; etc.)Considerable concentrations of pollution can be seen within the perimeter of two largest coastal cities - Lattaquia and Tartous. The environmental degradation is the least felt in the northern part of the region which is less densely populated.

Having in mind three main geographical parts of the region - the coastal plains, the hilly area, and the mountains - the major sources of degradation and environmental conflicts in the region are rooted in the concentration of activities and population in the narrow coastal plains. These are:

-   fresh water pollution and sea water intrusion in the ground water due to overpumping;

-   sea water pollution caused mainly by liquid waste discharges from big coastal cities;

-   damaging of landscape by quarrying (extraction of sand and pebbles), solid waste dumping and illegal housing;

-   encroachment of agricultural land, sites allocated to urban and tourism development and areas of special natural value by haphazard or illegal housing;

-   indiscriminate concentration of versatile activities in the narrow coastal strip.

Some major findings and assumptions of the Integrated Planning Study, which may indicate the course of the future development of the region, are the following:

-   All basic regional resources (agricultural land, economic activities, infrastructure facilities, large urban settlements, natural, landscape and cultural values) are located in a relatively narrow coastal zone.

-   The existing alternative strategies of development and population distribution indicate that migrations are likely to continue to flow towards the urban and rural areas of the coastal zone.

- The bulk of the future economic development expected in
  agriculture, manufacture and tourism (which mostly draws
  on the existing infrastructure facilities) will also be
  concentrated in the coastal zone of the region.

Compared with the rest of the region, and judging by the
existing state of the environment, natural resources of the
coastal zone are the most endangered ones (sea water, fresh
water, beaches and areas of natural value).

To avoid that the existing conflicts between the vital
development activities and the environment of the coastal
region persist in the future, a comprehensive planning has to
be developed and immediate actions undertaken to be the basis
for a careful management of coastal resources.

Generally speaking, the major objective of the two-year
programme (1990-1991) corresponds to the objective set for the
previous period, that is, to use the method of integrated
planning and management of resources to achieve a higher
degree of harmony between the development and environmental
protection of the Syrian coastal region. The knowledge
obtained through the preparation of the Study points out the
need to continue with the integrated approach to the
interdependent problems of the economic and demographic
development and the protection and promotion of the physical
environment in which these occur.

The strategy of the programme departs from the
assumptions discussed in the Integrated Planning Study and
further elaborate them. The stress has been shifted from a
general view of individual activities to precisely identified
problems and priorities, and through them, to the
implementation of general development concepts as proposed in
the Study. The selected problems and priorities, elaborated to
various degrees, have to follow the general concept of
integrated planning and management of resources.

Such a strategy requires a selective approach to the
identification of areas in which the majority of activities
will be carried out. Since the problems of pollution, as well
as development resources and potentials, are concentrated in
the immediate coastal zone of the region, the programme should
focus on that particular area.

The long-term objective of the programme is to protect
and rationally utilize the coastal resources over a relatively
long period of time. The task of such a programme is to
determine and recommend management measures (particularly in
the domains of land and sea use, environmental protection,
rehabilitation of historic monuments, etc.) with a view to
resolving the existing environmental conflicts and setting up
the optimum paths of the future dynamic development.

The main expected outputs of the programme are:

- proposals for immediate actions;

- technical and economic measures for addressing existing environmental problems;

- integrated management plans;

- studies and reports on specific subjects;

- training of local and national experts;

- demonstration projects;

- monitoring programme;

- data base for various development and environment aspects;

- software to be used in solving some specific problems.

The following activities are performed within the programme: implementation of the land based sources and the dumping protocols; implementation of the emergency protocol and the MARPOL Convention; monitoring programme of the marine environment; study on the implications of the expected climate changes; protection and management plan of historic settlements; training programme on and application of GIS on pcARC/INFO within the planning process; EIA of the Amrit tourism development project; development-environment scenario; coastal resources management plan; programme for the protection of specially protected areas.

Hitherto Achieved Results and Experience
Although all of the 4 described MAP CAMPs are still in the phase of implementation, it is already possible to present some experience and achieved results:

a) The importance was confirmed of a good and thorough preparation regarding the institutional arrangement:

- in 3 of the 4 programmes, there were delays and difficulties in the preparation and starting of the implementation due to causes that belong to this domain;

- in 3 of the 4 programmes, changes occurred of either national or local authorities, or both, in the preparatory or implementation phases; in 2 of the 3 cases, it caused postponing of the start or delays in the implementation of some of the activities, while in 1 case it caused a delay in the start of the entire programme.

b) With regard to a timely provision of a sufficient source of financing and it punctual availability, in 2 of the 4 cases difficulties were faced due to which, in one case,

148

it will not be possible to complete some activities in the envisaged time.

c)  A good preparation of <u>local teams</u> and creation of adequate conditions for their work, in 3 of the 4 cases, contributed significantly to the implementation of the programmes and compensated for some of the above mentioned difficulties.

d)  The importance was confirmed of the <u>training component</u> in the preparatory and implementation phases of the programme. Local teams with little previous experience, cooperating in an appropriate manner with MAP experts and consultants, implementation of activities such as had never been implemented in those areas and/or countries.

e)  A number of very important activities were performed, resulting in applicable documents, and instructions and recommendations for immediate and medium-term actions: integrated planning studies and documents, development scenarios, studies on the impacts of the expected climate changes, and results of monitoring and research into sources and trends of pollution. As pilot applications, EIA and GIS were successfully tested in the planning process.

f)  Local teams were trained, or are in the process of training, to prepare integrated plans and development scenarios, and to apply EIA and GIS.

g)  The basic objectives of the programmes are being met in a satisfactory way, while the delays remain within acceptable limits.

h)  By the end of 1991, in 2 of the 4 cases, 75-80% of activities was completed, and in the other 2 cases 50-60%. All of the 4 programmes will be completed in the envisaged terms: 2 by the end of 1992, and the other 2 by the end of 1993.

i)  In all 4 cases, the activities already performed and those realistically expected to be completed, represent a good basis for a future cooperation of MAP with the national and local governments in the field of coastal zone management, in the selected and other areas. The established, well trained, local teams, as well as the local authorities will be capable of continuing their activities, using the principles and methods of ICZM to a degree much higher than before.

j)  It can be expected that it will be especially the local authorities and institutions that will be interested in some form of follow-up to these programmes. That will probably be discussed in the MAP meetings in the 1992-93 biennium.

The results of individual activities within the on-going programmes have already been used when designing concrete projects. In the cases of the Izmir and the Kastela bays it refers to the projects of collection, treatment and disposal of urban wastewaters. They will also be used for defining submarine outfalls. In 3 of the 4 projects, the prepared planning documents will make the basis for the preparation and adoption of land-use, sea-use and urban plans (Syrian coast, Rhodes, Izmir).

For each MAP CAMP, upon their completion (during 1993), final reports will be prepared and evaluations made. The reports will be presented to the national and local authorities, and later put at the disposal of all Mediterranean countries.

Finally, it should be mentioned that the Seventh Ordinary Conference of the Contracting Parties (Cairo, October 1991) authorized the preparation and launch of 3 new MAP CAMPs to be implemented in the areas of Fuka (Egypt), Sfax (Tunisia) and Durres-Vlora (Albania). All the experience gained in the implementation of the four previously described on-going programmes will be used for the formulation, preparation and implementation of the new projects.

## REFERENCES

1.  van der Bergh, J.C.J.-M., and Nijkamp, P., Operationalizing sustainable development, Ecological Economics, 1991, 4, pp. 11-13.

2.  Sörensen, J.C., Institutional arrangement for managing coastal resources and environments, National Parks Service USDI and USAID, 1990.

3.  Coastal Areas Management and Planning Network, CAMPNET, Workshop "The Status of International Coastal Zone Management", Charleston, SC, USA, July 1989.

4.  The Priority Actions Programme Regional Activity Centre, MAP, UNEP, Methodological Framework for the Process of Integrated Planning and Management in the Mediterranean (PAP-4/EM.5/2), 1988.

5.  Grenon, M., and Batisse, M., Futures of the Mediterranean Basin, The Blue Plan, MAP, UNEP, 1988.

6.  An Approach to Environmental Impact Assessment for projects affecting the coastal and merine environment, UNEP Regional Seas Reports and Studies, 1990, 122.

7. PAP/RAC, Methodological Framework for Assessing Tourism Capacity in Mediterranean Coastal Zones (PAP-9/P.1), Split, 1990.

8. Mladineo, N., Multicriterional Analysis for Environmentally Sound Siting of Development Projects Applicable in the Mediterranean Region, Split, 1990.

9. UNEP, Report on the Joint Meeting of the Task Team on Implications of Climatic Changes in the Mediterranean and the Co-ordination of Task Teams for the Caribbean, South-East Pacific, East-Asian and South-East Asian Seas, (UNEP(OCA)/WG.2/25), Nairobi, 1988.

10. Pernetta, J.C., and Elder, D., Climate, Sea Level Rise and the Coastal Zone: Management and Planning for Global Changes, IUCN, Gland, 1990.

# FINAL RECOMMENDATIONS OF THE INTERNATIONAL MEETING

The participants, after three days of extensive exchange of opinion and fruitful discussions expressed their appreciation for the initiative of bringing together representatives from the countries participating in the Mediterranean and the Caribbean Regional Seas Programmes, the opportunity given for the first time to qualified representatives from 15 Caribbean countries, 17 Mediterranean countries, of the EEC and of a number of representatives from UNEP, the Caribbean Environment Programme, the Mediterranean Action Plan and its Regional Centres, to discuss their common problems in selected priority areas is in itsealf a major achievement of the meeting.

After the presentation of the introductory documents on the experiences of the two programmes and a discussion of the three selected priority areas, the participants agreed on the need and the usefulness of continuing the "Colombo '92" initiative regarding the cooperation between the two Regional Plans by promoting the following:

A) **ON WATER QUALITY - (Exchange of personnel, of experinece and training):**
- on REMPEC activities in the Mediterranean and the Clean Caribbean Cooperative of the oil industry, in the framework of IMO for combating oil pollution;
- on monitoring, data quality control, use of remote sensing, search for environmental quality indicators; national legislation;
- on land-based sources and particularly on pesticides (criteria relating to human health and the environment, tests on biodegradation and bioaccumulation, costs);
- on management of plastic waste (collection, biodegradable plastics, incineration), a subject of considerable interest to Caribbean fruit producing and exporting countries;
- on control of transport of toxic waste across frontiers, based on the Basel Convention and regional protocols;
- on cost-benefit analysis of investment for the abatement of marine water pollution, including sewage, port reception facilities and treatment of industrial effluents;

B) **ON THE SUBJECT OF COASTAL ZONE MANAGEMENT - ( Exchange of personnel, experience and training on):**
- methodology and practice of coastal zone planning and management in order to achieve a sustainable development;
- coastal resource economics and accounting;
- ecological zoning of coastal areas for the location of

productive activities;
- methodology and practice of conservation and rehabilitation of historic settlements;
- planning, development and management of tourism activities harmonized with the environment, including protected areas for the promotion of eco-tourism;
- identification and management of environmentally sensitive areas (coastal and marine parks, coral reef, mangrove forests);
- beach management (study of beach dynamics, beach cleaning, artificial beach nourishment);
- preparation of long term scenarios of environment development interrelations in the Caribbean Region based on the experience of the Blue Plan in the Mediterranean and of environmental land use planning in Mexico and Latin-American/broader Caribbean region (requiring a major investment in expertise, funds and time);
- preparation and dissemination of handbooks and guidelines on methodology, tools and techniques applicable in the coastal zone management process in both regions to manage the conflict between different economic activities;
- innovative financing instruments and legislation for the support of environment protection measures by the private sector using the coastal zone;
- use of national capabilities in order to develop the cooperation in the coastal areas.

C) **ON THE SUBJECT OF CLIMATE CHANGES** - (Exchange of personnel, experience and training on) :
- geomorphological studies of the coastline to assist in predicting the local effects of sea-level rise;
- national plans for protection measures in island-countries with limited land surface having no retreat alternatives;
- impact of volcanic activities on water quality and coastal management;
- national case studies with UNEP support;
- common applicable vulnerability indices for the preservation of coastal resources;
- models of beach tourism management;
- methodology for the evaluation of site- specific responses to sea level rise scenarios, in terms of realistic time scales and of the cost effectiveness of development.


The meeting also discussed and recommended the following actions of a general scheme to be promoted:
- promotion of subregional cooperation on the model of the RAMOGE, UMA, etc.;

- involvement of high level decision makers and managers though appropriate actions (meetings, information, field demonstrations, etc.);
- development of mechanisms to involve the private sector in the Mediterranean-Caribbean cooperation;
- promotion of Eureka-type projects for the joint development of technology between Mediterranean and Caribbean countries;
- development of training in environment techniques through exchange of professors and students (Erasmus) and stages in industry (Comett);
- awareness of cultural aspects relevant to the developement process and to the protection of the environment;
- dissemination of information to the local population and tourists;
- facilitation of access to data and information on coastal zone environment, land based sources of pollution, coastal zone planning and management;
- creation of a joint network of institutions and experts.

Additional finances and support - The meeting also invited:
- the EEC, the World Bank and international financial institutions as well as national Government to take into account in their future financing plans the new opportunities for training and exchange on the specific issues identified by the Meeting;
- developing as well as industrialized countries to give higher priority to programmes for environmental protection in their bilateral and multilateral cooperation;
- UNEP to continue its support for cooperation between the two Regional Seas Programmes;
- the City of Genoa with its academic and research institutions to continue its support to this "Colombo '92" initiative.